STARTING AND MANAGING YOUR OWN ENGINEERING PRACTICE

STARTING AND MANAGING YOUR OWN ENGINEERING PRACTICE

JOHN A. KUECKEN
Electromagnetic Engineering Development

VNR VAN NOSTRAND REINHOLD COMPANY

NEW YORK   CINCINNATI   ATLANTA   DALLAS   SAN FRANCISCO
LONDON   TORONTO   MELBOURNE

Van Nostrand Reinhold Company Regional Offices:
New York Cincinnati Atlanta Dallas San Francisco

Van Nostrand Reinhold Company International Offices:
London Toronto Melbourne

Library of Congress Catalog Card Number: 78-4686
ISBN : 0-442-24513-0

Manufactured in the United States of America

Published by Van Nostrand Reinhold Company
135 West 50th Street, New York, N. Y. 10020

Published simultaneously in Canada by Van Nostrand Reinhold Ltd.

15 14 13 12 11 10 9 8 7 6 5 4

Library of Congress Cataloging in Publication Data

Kuecken, John A
  Starting and managing your own engineering practice.

  Includes index.
  1. Consulting engineers. I. Title.
TA157.K83   658'.91'62   78-4686
ISBN 0-442-24513-0

Preface

It will come as a surprise to no one over the age of seven that people come in many different sizes, shapes and temperaments. Dandy Don Meredith (fearless QB for the Dallas Cowboys and TV commentator) says: "Different strokes for different folks." He is, of course, right.

If one considers mankind as one large pie, a number of classifications would divide the pie about in two. The most obvious of course is sex. Roughly half of mankind is female and the other half male. However, there are other, more subtle, binary divisions of the pie.

For example, one of my college professors maintained that most groups of people, if confronted with an ant farm (one of those glass gadgets in which the workings of an ant hive are exposed to view), would divide about equally into two parts: (1) *the passive*—those who would view the proceedings passively, out of either (a) boredom or (b) awe, marveling at the workings of a magnificently ordered society; and, (2) *the active*—those who would tap on the glass with a fingernail because this makes the ants run around (this provides a good deal of activity; it yields something to watch!). Try it for size! Does your office, church group or bowling league divide roughly half-and-half into watchers and glass-tappers? My groups do!

The reason for these comments is, of course, an entirely different binary division. Over the years I have become convinced that within the breast of roughly half of all men beats the heart of an entrepreneur. Among engineers, my estimate runs 71.3%. The human race is divided between those who would *like* to establish their own course and those who would prefer to be led or assisted. This book is naturally addressed to the first group. To them the thought of directing their own lives, independent of the organization, the boss or management, can sometimes be a siren song luring them away from the channel of their lives. THINK OF IT!

An end to those confounded grievance committee meetings with the pigheaded shop stewards!

No more arguments with the imbeciles from the Trust Fund and the SEC!

Joe Fotzengargle can beat someone else over the head with his idiotic ideas about how his company's disgusting cake mixes should be advertised by his execrable female STAR!

Sweet respite from those cretins who send back a perfectly good load of antennas because they don't know how to read VSWR on a reflectometer!

Fill in: _____

_____

_____

_____

_____

_____

(Attach additional sheets as required.)

Of this independent group, a few actually manage to do it. In the U.S. Virgin Islands, on a relatively short stay, you could meet a tobacconist who was once director of labor relations for an auto maker; a charter boat captain who was once an account executive for a major brokerage firm; a glass-bottom-boat skipper who was once chief salesman for a sizable broadcasting network. And on rare occasions you might run into a self-employed engineer who sometimes writes books. He is there enjoying the hospitality of a brother-in-law who left a big-city tax law practice for a general law practice amid the golden sands in that balmy clime.

Lest the siren song become too sweet too soon, I will note that the tobacconist still has to run a business, the charter boat captain had to obtain U.S. Coast Guard Captain's Papers, the glass-bottom-boat skipper had to buy the boat and business, and the attorney had to spend a year in residence, working at other jobs, before he was even entitled to retake the bar exam. The engineer, alas, gets to go reef diving pretty rarely because of the cost of air fare! The point is that it can be done *if you are willing to pay the price*.

However, for an overwhelming majority, the dream will remain a dream forever. The pension plan is too good to ignore! The hospitalization plan would be too expensive to duplicate! The kids' education is too costly to permit that kind of a gamble at this time! We would have to give up The Country Club, The Summers at the Lake, My Airplane, The Camper, My Hairdresser, Skiing and a million-and-one other items. Each of these is a good and valid reason for not casting aside the fetters.

In addition, it is well to contemplate the thought that not all who make the leap survive. A new 43 H.P. diesel auxiliary for your charter ketch will cost six BIG ONES! Let Uncle get friendly with Castro and your $10,000 investment in Honduran Cigars is out the window. The project is a natural for a microprocessor, but I am looking straight-down-the-barrel at $1,600 for a teletype terminal, $6,800 for a logic analyzer, $4,950 for an editor/compiler and $7,120 for a video terminal. THERE IS NO ONE ELSE TO PICK UP THE TAB! "On your own" means *just that!*

In addition, a few surprises await the unwary. Take Social Security, for example. When you are working for a firm, the firm picks up about half. When you go it alone, you will discover the wonders of the "Self-Employment Tax." *You* get to pay a far greater amount than before. Unemployment insurance comes as something of a revelation too. If the whole thing doesn't work, are you eligible to draw unemployment compensation while you try to find something else? Check your state law!

There are other revelations for the individual operating a private engineering practice. For example "net 30 days" does not always mean that you will have your money 30 days after your invoice is sent, even when the client is well funded and well run. Your assistance was needed in a hurry; however, the fellow who has to approve the transfer of funds may be on an extended business trip or a vacation. The manager who wrote the purchase order may have charged it to the wrong account. Accounts Payable or Legal may send the PO back to the originator for clarification. There are a thousand and one reasons why a perfectly good client may be tardy in settling a bill. In the meantime, your bill for airplane tickets, rental car and motel room arrive, and *you* get to pay in full or handle the funds on a 1.5% per month carrying charge.

The object of this discourse is not to discourage, but rather to point up some of the very real storms to be weathered in the establishment of a private engineering practice. It is not something to be done easily or lightly; however, it can be immensely rewarding to the individual. I hope to have something useful to say about each of these difficulties, and intend to offer suggestions for optimization of survival probability. If you pay no more for the lessons than the price of this book, you will get by more cheaply than I did!

If you manage to enjoy the private practice of engineering as much as I have, the effort will be well worthwhile.

JOHN A. KUECKEN

Contents

tions applied to specific cases. A series of tables presents the monthly income, budget, tax and savings through the formative years. It is shown that a very narrow "window" for success exists with a nearly unavoidable crisis at the three-year point. The income is shown to approximate the pre-practice level at about five years. The savings or initial investment is shown to be restored at about the eight-year mark.

STARTING AND MANAGING YOUR OWN ENGINEERING PRACTICE

1

The Practice

First on our agenda is an attempt to define "the practice." Used as a verb, the word practice means to perform an act repeatedly in order to obtain skill or proficiency; as in: practicing scales on the piano or violin, or practicing a tennis or golf swing. This meaning is not far off the mark for our purposes, but we should look further.

The noun practice usually refers to the clientele of a physician or an attorney; the patients or clients who supply the livelihood of the practicioner, his source of income.

From earliest times, the physician and the attorney have been admired by their contemporaries. They are self-determined men. They make their own way in life. Kings, captains and peasants alike come to them for balm, comfort, help, counsel or succor in time of trouble. They inspire the flame of admiration in all men. Among engineers, this flame seems to burn especially bright. In the meetings and the professional publications of the IEEE (the Institute of Electrical and Electronic Engineers) and the NSPE (the National Society of Professional Engineers) one hears the continual chant calling for "professional recognition." The engineers want to be recognized as "professionals." And they are not alone in this desire. Hairdressers, floor-care specialists, dental technicians and guidance counselors also want "professional recognition." Each group has had some special training, and each has learned an art. They have each been educated in something. Now what is it that defines a "profession"?

The term "profession" as distinguished from "job" or "trade" is always used in connection with practitioners of medicine and law. It is also applied to "the world's oldest profession." These practitioners provide personal service to individuals for profit; but then, so do the barber, the beautician and the auto mechanic, and few consider their occupations professions. So the "personal service for profit" definition doesn't work too well. What other criteria have we to apply to distinguish the mere "job" from a profession?

Well, the clergy and educators are often included in "the professions." However, the clergy are generally supported by a parish council, a board of elders or a bishop, while educators are generally supported by a board of education or a board of regents (sometimes on a contract negotiated by the International Brotherhood of Educational Workers/Division of the Teamsters Union). They do not exactly fit the self-employed or self-determined definition.

However, there is a common denominator. The physician, the lawyer, the clergyman, the teacher and, alas, the prostitute have devoted a considerable amount of time and effort to learning their art. But, unfortunately, this definition also crumbles when one considers the deep-sea diver, the watchmaker, the olympic skater and the crane operator, who have devoted equivalent time and effort. Surely they are not professionals. The olympic skater would be disqualified!

Maybe there is a clue to be found in an examination of what the engineering organizations are asking for. At the time of this writing one of the loudest cries is for job security and seniority, or, more politely, "tenure."

During the cutbacks in the aerospace industry in the late sixties and the electronics industry in the early seventies, many engineers were laid off. Since the engineering groups are usually not unionized, the lay-offs were generally not done strictly on the basis of seniority, although it was sometimes considered. When the time came to rehire, the senior engineers were sometimes replaced with new graduates, who were cheaper and presumably more up-to-date. Thus we hear the cry in the professional societies for job security and seniority. This desire may be pragmatic; however, it is not distinctly professional!

Another cry is for portable pensions. The Teamsters have had them for a long time!

Another cry is that senior engineers have been rehired at new-graduate rates and that the professional societies should impose sanctions against such employers. Collective bargaining may be profitable, as demonstrated by the electricians', plumbers' and auto workers' unions; but it is not distinctly professional.

Thus we see that none of the definitions really fits very well. I will therefore submit the definition that a professional is some-one who is considered by his neighbors to be "a professional." It is not something that can be legislated or directed from the out-side; so don't worry about it!

However, this discussion does have something to do with "the practice" because it contains some fundamental arguments implicit in the description. The first of these deals with the nature of the clientele. Whereas a physician or an attorney may have a clientele made up entirely of individuals, a Certified Public Accountant will only rarely have a practice composed entirely of individuals. A Consulting Engineer, on the other hand, will rarely if ever have even a single individual as a client (with some exceptions).

Let's talk about that for a bit. The question of whom one is to serve is a significant consideration. There are few if any of us who have not had reason to call on a physician for help. He delivers our children and treats them when they have measles, chicken pox, flu or whatever. The physician who came when my middle daughter was delerious with measles calmed *me*, as well as treat-ing my daughter. She inspired my confidence. She handled the situation.

In the area of the law, for most of us, the association is a little less dramatic. We write a will, buy a home, sell property, some-times sue someone and occasionally apply for a patent or want to sue someone for infringing on one that we own. On the other hand a few of us have to bail out Uncle Otto for D.W.I. or nephew Herman for Criminal Possession. The attorney serves us strongly and directly in such cases. If the task is well done, we appreciate his skill and efforts.

On the other hand, only a relatively few very wealthy individuals would have any reason to retain a Certified Public Accountant for personal matters. The art of the CPA is almost inextricably associated with business and commerce. The rather sizable number of individuals who do retain the services of a CPA do so for business reasons.

The consulting engineer provides a still stronger case. The civil architect or the marine architect might be retained to design a home or a yacht for a very wealthy individual, but even this service is a rarity rather than a rule. Far more of their time is spent on public buildings, churches, schools, tankers and destroyers. Rising labor costs and taxes have made the disparity greater each year, and the number of clients with great disposable wealth continues to diminish.

For the mechanical, civil, aeronautical, chemical, automotive or electrical engineer with a consulting practice, the chances of obtaining a private client for a nonbusiness assignment are vanishingly small.

The reason for dwelling upon the distinction between the types of practices is that this point reveals a great deal to us about the nature of what must go on.

First: If your main reason for going into private practice is to avoid or shed the problems of the business world, FORGET IT! The consulting engineer or CPA or broker must not only run his own business, he must continually deal with the problems of other people's businesses. Their problems are your problems, or they won't be your clients very long.

Second: If you expect that going into private practice will enhance your "professional recognition," you can forget that one too. No amount of success in dealing with proprietorships, partnerships or corporations is going to make your neighbors look upon you in the same way they look upon the obstetrician who delivered their children or the attorney who saved their home from foreclosure. They will neither know about nor understand your successes.

Third: Before you start, you had better work out a pretty good picture of just who your clientele will be. If you cannot identify enough classes of business as logical potential clients to give you more work than you could possibly handle, you might want to rethink the situation before turning in your resignation. In this consideration it is especially important that you be brutally honest about why it would be to the advantage of the client to engage you. He will not do it because of your charm and beauty. In general he will only do it to make or save money.

REASONS TO RETAIN A CONSULTANT

A variety of valid reasons might prompt a businessman to retain a consultant. These reasons will be discussed at length later, but we might look at them here, as they reflect upon the nature of the practice.

Perhaps first and foremost is the case in which the businessman has a limited requirement for a high skill level in an art not represented in his firm. For example, a real estate developer generally needs the services of a civil engineering firm to lay out and specify sewers, drainage, roads, etc. When the development is finished, the requirement for that skill does not exist within the real estate firm. A similar argument applies to the architect who designs the buildings. This is one of the largest markets for consulting services.

A bank can have a similar requirement for selecting, assembling, testing and writing initial software for a computer system. Once it is up and running, a few programmers and perhaps a few technicians are all that could be economically justified.

In these cases, the client obtains high-level services for a limited task without permanently enlarging the payroll.

The second most prominent reason is in the area of problem solving. In manufacturing, from time to time pernicious and baffling problems will arise. They can often be costly or even disrupt the normal work flow. Top management will often direct those responsible to "Get in an Expert." A consultant with a good reputation for problem solving and expertise in the area will be engaged. This type of consulting work is reasonably profitable; however, it has several drawbacks not represented in some other types.

1. The problem may be very difficult to solve within the practical and financial constraints of the situation. Remember, they tried pretty hard before they called you.
2. The work is sporadic in nature. No one plans in advance to have a problem. The call will come in, and immediate response is required. The unplanned nature is less a problem for the college professor who does a bit of consulting as a sideline than for the full-time consultant who must strive for a reasonably smooth work flow.
3. These latter tasks are always smaller than the first kind. If the problem stops production, you will either solve it in a hurry or you will fail. Too many failures or not-soon-enough solutions will hurt your reputation for this type of work.

The third most prominent reason for retaining a consultant is cost effectiveness. This applies particularly to large corporations involved in the development of a new product, particularly when that new product differs significantly from the existing product line. The firm may have the required skills represented but may be inexperienced in specific techniques. A consultant with experience in these techniques can often arrive at the new product enough faster than the in-house group to make the costs significantly lower.

A good example of the above is the recent trend toward all-digital devices in the instrumentation industry. As recently as 1970, the instruments used to monitor the operation of oil refineries, bakeries and soap factories were analog instruments. They operated by varying the current flow in a circuit, which, in turn, would move the needle on a remote indicator or the pen on a chart recorder. This stepless proportional control of the remote instrument works reasonably well, but the accuracy is limited. One can read the needle position to an accuracy of 2 to 5%. The trend among the instrument makers is away from analog (or proportional) instruments and toward digital instruments. This change offers several advantages in that the display can be made to read to any desired accuracy, and the output of the instrument can be directly fed into a computer to automate the process being controlled.

The instrument makers were well staffed with electrical engineers; however, those engineers were generally not experienced with the digital techniques, which are very different from the older analog techniques. The in-house engineers would have to have some time to experiment, make mistakes and learn. This can be an expensive process.

One obvious solution would be to hire some new people who had digital experience. This route was taken by quite a few firms. The alternative was to engage a consultant to develop the first or the first few of the new instruments. The in-house people could then learn by working on the new line, and a smooth transition could be obtained. By the time that the first couple of instruments were in production, the services of the consultant would no longer be required.

Sometimes, when the change is really large, one firm will contract with another firm to develop and build the new product line. This was the case with several of the old-line mechanical calculator and adding machine makers. Their staffs were expert at designing and building precision gear trains, mechanical latches and keyboards. The electronic calculators were so much faster, quieter and simpler to construct that it was obvious that they would rapidly come to dominate the market. However, the techniques involved were so different that it was questionable whether any significant fraction of the work force could be retrained.

A number of these firms contracted with new, small electronic computer/calculator houses to develop and build the units with the label of the contracting firm on the units. In a number of cases where this scheme worked out well, the small firm was eventually purchased by the larger firm. The mechanical calculator fabrication group was allowed to wither on the vine, supplying spare parts as long as a demand existed. For the parent firm the marketing and service divisions were forced to undergo some changes, but the firm itself survived.

This topic will be discussed further. It is not exactly consulting engineering, but it illustrates the use of outside help in a product-change situation.

One point should be noted from this discussion. The consultant is of value in the product-change situation only as long as he is well ahead of the art. Once a given technique is established, his economic value in the situation ceases. In effect, he eliminates his position. Obviously, if he is to survive, he must be ahead of something else by that time. This can be quite a footrace.

A fourth reason for retaining the services of an independent consultant I will describe as prestige, although it requires a little explanation. It has to do with the use of the outside consultant because of his name and reputation.

To cite a specific example, I retained one of the "Big-Eight" accounting firms to set up and handle my books. In actual practice, on a month-to-month basis the books are handled by our esteemed treasurer (my wife). At the end of the year when all the returns are in, she balances the books and then we call up the accounting firm. We get together with the CPA assigned to my small account, who closes and audits the books, and prepares the state and local tax forms and the estimated tax for the next year. This is a typical CPA function.

The same tasks could have been performed in-house if I were smart enough. I could also have gone to one of the store-front tax-preparing firms or could have retained our friendly neighborhood CPA, who audits the books for the filling station and the bowling alley. Instead, I elected to pay a premium for a "Big-Eight" CPA who specializes in small business and medical and legal practices.

I think I save money in the long run because my practice is characterized by many peculiar entries. Travel expenses are often the largest single expense item. Receipts of a significant fraction of total income can be in foreign currencies. There are expenses for currency exchange, etc. By having a "Big-Eight" audit I know that the CPA who did the work was qualified and had the resources of experts at his disposal. Furthermore the tax reviewer and my banker know it! The imprimatur of the firm on these matters at least gives *me* great comfort.

Probably the best example of this factor is to be found in the practice of having an outside firm audit the books of a corporation before the stockholders' meeting and tax time. The outside auditor is of value precisely because he is an outsider. He presumably would not profit by any hanky-panky with the corporate funds and can function effectively only so long as his reputation implies scrupulous accuracy and attention to detail.

This last procedure is so common that a series of very large businesses which do nothing but accounting have developed in the United States. I estimate that the "Big-Eight" accounting firms represent the very largest businesses that could be described as a practice. In such a case the firm itself would be described as having the practice, while the individual CPAs would be employees, not in private practice. These firms are larger by an order of magnitude than anything else in the consulting line.

The next largest firms are the various not-for-profit research foundations such as Battelle, Stanford Research Institute, and so on. They are big businesses also. They do indeed perform a consulting function and may be involved in any of the three types of consultation previously discussed. However, to the individual engineer they represent employment and not private practice in the sense described here. The prestige of these firms is frequently used to support a public or legal position.

For the individual in private practice, this last-mentioned form of commission (support of a position) can arise in smaller matters. Sometimes it will arise in courtroom proceedings, and the individual will be called upon to testify. This is an interesting situation, in that the Technical Expert is the only witness allowed to testify on a matter of his opinion, within the area of his expertise. In such matters it does not hurt to be the head of a department at MIT or Harvard or Cal Tech.

Not all of the commissions of this kind are as important as a major trial or a congressional hearing. On occasion the consultant will represent a firm at meetings and conferences and before the firm's customers. If you are not a department head at a major university, this is about as far as you can usually go in this category.

The accountants have a clear edge in this type of prestige representation and the engineers are the least likely to find much in this department. In general, those commissions that do arise are brief, but usually enjoyable.

THE PRACTICE VERSUS THE SMALL BUSINESS

Whenever the art is such that some physical product can result, as in engineering, one frequently finds that the practitioner runs a small business as well as doing the consulting. For example, the man who is an expert on flugel pins may have a small machine shop in which he manufactures limited runs of very special-purpose flugel pins. He acts as a consultant to the flugel pin industry as well. We shall be discussing this at length in the next chapter, but here we should note that the existence of the flugel pin small business can very effectively shut him off from major segments of the consulting business. No one would really want to turn over major product development or production to a potential competitor. As a result, those who succeed in this type of arrangement tend to show a pattern of less and less consulting practice and more and more small business. As the small business grows, it shuts the practitioner off from progressively larger numbers of clients. This is great if you want it that way.

There is a distinction to be observed concerning the type of hardware that is generated. In a hardware business there is no substitute for building and handling hardware. If the consultant in the previous example were retained to develop a new type of flugel pin, and he built and tested a variety of them that were subsequently delivered as prototypes or samples to the client, this would displease no one. He would in no sense be competing with the client for any portion of the flugel pin market. It is extremely important that the competitive position of the client be protected, especially from you. If you wish to, or must, operate a small business in addition to the consulting practice, it is well to operate in a nonoverlapping area.

For the person who expects to do some inventing or development work in his practice, the requirement for some physical development facilities is imperative. It is a physical world that we live in, and a "good idea" is never really a good idea until it has been made to work. One can reason and calculate and theorize ad infinitum about some gadget; however, it is not until its actual operation has been proved that one has done anything of significance.

As will be shown in subsequent chapters, the operation of a particular principle is never really convincing to a client without a physical demonstration of some kind. Even in matters as non-physical as computer software, it is the rare exception if a new program runs the first time exactly as written. There is always a de-bug procedure included in software development programs because the people who regularly do this sort of work are aware of the fact that there will be errors and glitches in the initial program. A software program is not a program until it has been run and proved on a computer. Also the fact that it runs on one computer does not necessarily mean that it will run on every computer.

In the hardware-oriented fields this is even more true. An idea that seems to be simple and straightforward will sometimes prove to be entirely inapplicable for some entirely unforeseen reason. The reason may be discernible only after the idea is tried.

For example, a number of years ago I was doing a series of studies on the characteristics of packset radio antennas in the high frequency band. On a packset radio the largest antenna that a soldier can carry while still functioning in a reasonably military manner is about 6 feet or 2 meters in height. At the low end of the military HF band near 2 MHz the wavelength is 150 meters; therefore the antenna is only 1/75 of a wavelength tall. This size is very tiny electrically, and it makes the antenna very difficult to drive. After a program of measurements I discovered that the antenna was actually radiating less than 3% of the energy, while the remaining energy was dividing roughly 75% into the soldier's body and 22% into the electrical network which attempted to match the antenna system impedance.

This led to further work to improve the military radio antenna, in which I discovered that there were large differences in the split of the power between the network and the soldier. These seemed to be correlated with the body weight or more properly with whether the soldier was fat or thin. A special instrument was then constructed, and very good correlation was obtained between the electrical properties of the body and the fat-thin relationship. It struck me that this was a discovery of considerable importance to the meat-cutting industry. Imagine, an instrument that would tell whether an animal was fat or lean without injury in a nearly in-stantaneous reading! It seemed like a natural.

I took my instrument to various packing houses and attempted to demonstrate it. This lead to three discoveries:

1. The stability of my digestive tract was not quite as great as I had been led to believe.
2. On live animals, this estimate could readily be done by eye-ball by experienced practitioners as fast as they could read the instrument.
3. This reading was of interest in detail only in the sausage-making industry where it is regulated by the Food and Drug Administration.

In the sausage-making industry the percentage of moisture in the sausage is controlled by the fat/lean ratio of the meats which are put in the sausage and a certain amount of process water that is added. The moisture is determined by carefully weighing a sample of the sausage filling and then drying this sample under a heat lamp and weighing it a second time. The weight loss divided by the original weight is taken as the moisture fraction. The fat con-tent is taken by weighing a sample and then cooking out the fat, which is poured off and weighed. I was confident that my instru-ment would find a valuable nitch in the sausage-packing industry. The existing process was messy and time-consuming, and an in-stant electronic read-out would be a considerable help.

Accordingly, I located a cooperative sausage packer, who agreed to participate with me in accumulating data that could correlate the instrument readings with the measurements of fat and mois-ture content obtained conventionally. The measurements were made by the lab at the sausage plant as a normal part of their qual-ity control procedure. In addition, the food chemist would take readings of the same sample with my instrument. The data were recorded together for interpretation and correlation to obtain a calibration of the instrument.

The first set of data covering a reasonably large set of samples was examined and showed no sign of the good fat/lean correlation I had obtained on the soldiers. After some consideration of this problem, it occurred to me that I was measuring only a part of the electrical parameters. The instrument was redesigned and arranged to measure *all* of the electrical parameters, and a new data gathering experiment was planned. The results were just as jumbled. The instrument was reading something; however, there was no way to relate the readings unambiguously to either fat or moisture content. All this despite the fact that I was able to get what seemed to be very good fat/lean readings on living soldiers.

A discussion with a biologist finally pointed out what may have been the source of the disparity. In a living animal, the percentage of salt is a precisely determined parameter, varying only within narrow limits; the moisture content is a function of the fat present and the flesh/muscle ratio. In the sausage, salt was added to the mix to the taste of the sausage maker. As it happens, the instrument that I had developed was particularly sensitive to salt water in the sample, and this was probably what the readings represented.

This may or may not have been the explanation; however, the point remains that I was never able to obtain usable readings on sausage. The program was abandoned from the meat-packing point of view because a reasonable resolution did not seem to be in sight.

Now the point of the matter is that a resolution of the question of whether or not one could make readings of moisture and fat content with the instrument could probably not have been obtained without some actual experiments and without building the instrument. The result was negative; however, I at least learned that it was. In the absence of the physical experiment I would probably still be convinced that it could be done.

It is important to note that an unsuccessful experiment does not ever prove that something cannot be done. It can only prove that that experiment did or did not do it. Another approach might very easily be successful. I simply haven't had the right idea.

THE PRIVATE PRACTICE SYNDROME

This chapter cannot be closed without some discussion of what I choose to call "the private practice syndrome." Whenever someone asks an engineer where he works and receives the reply "I am a consulting engineer in private practice," the phrase translates in the ear of the listener to "out-of-work." This is a remarkably stable phenomenon with a half-life of about three years.   Three years after one starts, half of his friends and neighbors will still say, "You haven't found anything yet?" or "I understand that Xerox is picking up Contract Engineers."   After six years, only 25% will sympathize, while the remaining three quarters will have become convinced that either you or your spouse inherited money.   By the six-year mark, you should have enough *clients* who believe that you really are in private practice to offset the effects of this attitude, and that is what counts!

2

How Do You Start A Practice?

At this juncture it seems logical to ask how one would go about starting a practice in engineering or some technological art. It would be a vast oversimplification simply to say that one quits his job and hangs up a shingle. In the final analysis, that is what it comes to. However, a great many steps must precede that action in order to make the effort succeed.

There is probably no single pattern of preparation that can be recommended as a sure-fire road to success in the consulting business. I have seen diverse circumstances lead men into the consulting field and be followed by success. The paths also vary with the particular field.

For example, in fields like marine or civil architecture or civil engineering there is little question that the most logical route is to spend a number of years working for a consulting engineering firm in the field. These fields are largely dominated by consulting engineering firms, since the clients would seldom have an economic justification for keeping that kind of talent on the payroll continuously. It simply makes better economic sense to contract for the talent when it is required and pay for it on a per-job basis.

In these particular areas, it is highly desirable (perhaps mandatory) that one obtain the requisite professional engineering license or certificate, since the function of these firms is concerned with public health and safety. In most cases you can work for such a firm without the license; however, some member of the firm generally has to have the license and must sign and approve drawings. If it is your desire to operate such a firm eventually, acquisition of the license is mandatory.

In such firms it is not unusual to find that the senior member of the firm may simply have inherited it from the person or persons who were his employers. In any event, the period spent learning the practice can also be used to acquire the license. If it does not seem to be in the cards for you eventually to inherit the firm because others are ahead of you or the boss has a son or daughter who is also qualified, you may eventually have to strike out on your own. The years of preparation will have trained you for the operation of the business, and you will know the techniques used to acquire clients and to handle the business.

In the case of the technologist—the mechanical, electrical, aero or chemical engineer—the matter is somewhat different. As noted earlier, the overwhelming majority of the engineers in these fields are employees of some corporation, and most of the work in their specialty is purchased by a corporation. The overwhelming majority of the engineering work in your specialty will be done on a regular basis by the employees of the firm. Unlike the case of the architect or civil engineer, the firm has a very good economic justification for maintaining a staff of engineers to handle its regular work. It is only in the special case that the corporation will bring in a consultant. For this reason, the form of the practice in such technological consulting is necessarily different from that of the architect or civil engineer.

In addition, the number of firms doing technological consulting work is very limited, and they are generally small, employing very few engineers compared to the industry as a whole. Thus there is very little opportunity for the would-be consultant in these fields to obtain employment from a consulting firm and learn the ropes by the apprenticeship route. Chances are that you will have to make your preparations and learn the lessons on your own.

Attached to every broad branch of engineering there always are little twigs of specialty. It is among these twigs that the consulting engineer flourishes. For example, among electronic engineers the antenna and propagation man is considered to be something of a specialist. To most electronic engineers, antennas are considered to represent an occult art. The possession of a specialty in one of these areas is obviously an advantage to the consulting engineer, since a great many of the firms in the industry will not have much need to support this specialty full time.

A similar argument for the specialist can be made for the person skilled and knowledgeable in EMI (electromagnetic interference) measurement and control. The requirement for this skill is relatively limited in most firms. The thought that the firm would go outside for such help is not at all distressing for most engineering managers.

The computer has fostered the development of an entirely new type of specialist, often found as a consultant—the software consultant. A discussion of his function serves to give a good example of the genesis and success of a market for a particular consulting function. A good deal of the story is old hat to digital electronic engineers; however, I ask their forbearance in deference to the mechanical, chemical and civil engineering readers of this book.

In the early days of the computer, it was a very large and expensive machine, which had many vacuum tubes, consumed a great deal of power, was installed in a specially protected room sealed off from the atmosphere and was provided with a private air-conditioning system. It was justified economically for the handling of bookkeeping chores and the preparation of the payroll. On occasion, the engineers of the firm were allowed to approach the priest who served this behemoth with a complex engineering problem to be performed between its financial duties. This man would plug a very large number of jumper wires between holes in what looked like a large square cribbage board, and *might* sometime later appear with a plot of your antenna pattern or the contour of the reflector dish duly typed out on wide sheets of green and white paper with rows of holes along the edges. These machines evolved fairly rapidly until, by the mid 1960s, an engineer could go down to the computer facility and turn in a great deck of punched cards with his problem written in a "High Level Language" and obtain his answers several days later in printed form.

As medium-scale and large-scale integrated circuits began to appear on the market in the late 1960s, a new form of machine began to make its presence known—the minicomputer. The minicomputer was not just a scaled-down big computer. Its significance lay in the fact that it was small enough and inexpensive enough to be practical to use for something other than handling the entire payroll and all of the engineering computations for a large plant. The minicomputer could be taken away from the data-handling province and dedicated to such tasks as operating one or more numerically controlled machines or controlling an automated chemical process. Compared to the early large central computer, this was an entirely different form of machine. It often did little or no arithmetic in the ordinary sense of the word.

In 1972 Intel brought out the MCS-4 Microprocessor as a vehicle to sell its solid state memories. Although this form of integrated-computer-on-a-chip could be used to build a tiny, slow computer for arithmetic and data-handling functions, the real appeal of the device lay in the direction of employing the unit to replace large assemblies of "hard wired logic" in automated controls and instruments. By the late 1960s, small- and medium-scale integrated circuits had developed to the point where engineers were busily involved in developing digital logic arrays that could automate nearly any process or measurement procedure, provided only that one had enough room and money.

As an example, the equipment developed to check out each of the electronic systems in the F-111 fighter plane contained more than 9,000 integrated circuit "bugs," probably about a half million transistors! I personally developed an instrument in 1973 and 1974 that contained nearly 500 IC packages.

The huge trailer-full of AGE (aerospace ground equipment) for the F-111 was merely an automatic sequencing and recording device. It was equipped to provide a stimulus to the airplane and then measure the response. For example, it would turn on a signal generator and then measure the output of the radio receiver. It was arranged so that an unsatisfactory response would change the nature of the testing as well. When the sequence was complete, the unit would type out a report showing deviations from the standards for the airplane equipment with perhaps suggestions for maintenance. In the ordinary sense of the word, there was no "language" for the machine and little if any computation, in the arithmetic sense of the word.

In my much smaller instrument, the device was intended to measure the frequency of an oscillator whose frequency was proportional to atmospheric pressure in very precise manner. The unit would then make corrections for the "law" or pressure/frequency relationship of the oscillator and display the result to six significant figures in millibars. It would remember the data on a decaying time base, and at the next reading it would compute the rate at which the pressure was rising or falling. The unit would print out the pressure, dp/dt and label the readings 32 times per hour. Once an hour it would also type out the time of day and the date.

The real lure of the microprocessor lay in the fact that potentially it could be "taught" to fulfill a large portion of these functions with only a few chips. For example, the 500-chip instrument could probably be constructed today using no more than 30 chips if one of them were a microprocessor. The entire sequence of operation could be burned into a ROM (read only memory), and the external sequencing of the machine would be indistinguishable from the earlier complicated one.

The principal problem with the use of the microprocessor is the fact that it looks at the world differently from the "hard wired" logic gate arrays that it replaces. A narrow 4-bit machine such as the MCS-4 must approach its problems a nibble at a time, performing tasks in sequence rather than broadside. The development of the sequence of instructions required to cause the machine to perform the desired functions is called "software development," and the program itself is referred to as "software" if it is stored in volatile memory which must be rewritten each time the machine is shut down, and "firmware" if it is stored in unalterable ROM. This work is very different in nature from the development of the large "hard wired" logic arrays, and it is typically done by punching a keyboard and observing a display rather than with a soldering iron and an oscilloscope. As of 1977, an overwhelming majority of electrical engineers had not yet had much contact with this art, which produces a printed string of hexadecimal characters for incorporation into the control memory. It is the province of the software consultant. With his expertise he can quickly derive the correct machine-language code for your application and perhaps teach the technique to your employees as well.

This work is nearly always done in machine language (which is nearly incomprehensible to the newcomer) for purely economic reasons. As an example, the code required to operate a series of three sets of traffic lights with such bells and whistles as time-dependent switching rates, traffic level adjustments and advanced turn signals can be written in about 128 bytes of memory in machine language. A truncated version of BASIC requires 8 to 16 K Bytes before any program is added. If it is used in any quantity, the economic advantage obviously goes to the single-chip machine language controller. The unique skill of turning out tight working code is obviously very much in demand these days.

Perhaps one of the most esoteric specialties is that of the EMP specialist, a person who is an expert in predicting the magnitude of the electromagnetic pulse that is radiated by the expanding fireball of a nuclear explosion. The protection of military electronic and power equipment from destructive electric arcs after a nearby nuclear blast is very important, and EMP hardening is frequently called out in military contracts. Instead of having such a person on their staff, the equipment manufacturers nearly always call in a consultant for this sort of effort.

In mechanical engineering, a number of specialties similarly lend themselves to consultants. The design of hydraulic servomechanisms is one such area. Carburetors for internal combustion engines are another. The practitioners of these specialties are not as rare as the EMP specialist; however, the skills are just far enough from the normal range of mechanical engineering that a great many firms may want the assistance of a consultant to augment the skills of its own staff.

My point here is simply that you can augment the opportunities for your success in the consulting practice if you are able to master one of the specialties of your field. You will increase the chances that a client will call on your services if you have some skill that is widely needed from time to time but not widely practiced.

Make no mistake, I am *not* counseling the would-be consulting engineer to learn some narrow specialty *to the exclusion of all else*. There is no hedge against financial disaster, either in the employ of a large corporation or in a private practice, that will ever match a broadbased competence in your profession. I am simply noting that the mastery of one or more specialties will bring you a significant level of extra business.

There is a second factor involved in the matter of having some expertise in a particular specialty. The company that retains you may actually want to do so because of your broad-based knowledge and reputation for engineering. However, if you have a reputation for expertise in some rather specialized area in which the average engineer is not expected to be versed, the entire process can be more easily explained and less embarrassing than it would be otherwise.

This is a more important factor than a first consideration might reveal. The retention of a consultant has a tendency to be a bit embarrassing in most cases. After all, your client probably has a substantial staff of engineers. Like most human beings, these people have a great deal of pride in their position. When the management calls in a consultant, it is implying that the task would not be successfully completed without him. At the very least, it implies that the task might go faster or more cheaply with his help. Unless the matter has been very delicately handled, at least some of the employees will interpret this as a vote of no confidence. It is very helpful to the consultant if he has some special qualification that is neither common to nor expected of the average engineer. With this, the employees of the client can accept the situation easily. After all, they are not being *paid* to be an expert in ——.

This brings up another topic. Just who is an expert? It has been observed that an expert is an ordinary fellow more than 25 miles from home who owns a briefcase. This might be close, but it doesn't quite hit the mark. If you want to be a consultant in some particular area, it doesn't hurt a bit if your fellows consider you to be an Expert. Also, we have used the phrase "expertise" in terms of the talents that you should be developing. Is there a definitive description of an Expert?

To define the Expert, let us start at the top and back down to see whether we can arrive at a definition of just who is an expert in any given field. Let us put the question on a personal basis: Just whom would you *have* to admit was an expert in your field? I submit that a man who has won the Nobel Prize in the field would *have* to be admitted as an expert by almost anyone! Okay, but there are very few of those! How about someone who is a Professor of the Topic at MIT, Cal Tech or Heidelburg? Well, you would probably have to accept him as an expert also! Suppose that he had written a book on the topic? How about 12 technical papers on the topic? A note in a professional journal? A Ph.D. thesis?

I submit that the judgment that a man is an Expert, like the judgment that he is professional, is something that is made by his neighbors. You cannot really influence, or at least legislate, it in the slightest. YOU *ARE* AN EXPERT IF YOUR CLIENTS THINK YOU ARE!

Do not mistake me! The time spent in earning a Ph.D. and writing 12 popular papers and a textbook on your topic are certainly well worth the effort. A half dozen patents in the subject area will not hurt either. However, when one client tells another that you are an Expert, that is worth more than all of the other accolades that you could garner put together! The point of this discussion is the fact that the would-be consulting engineer had better do at least one, or preferably several, of these things if he would establish some reputation for expertise in a given field.

PUBLICATIONS

Particularly important is the subject of scientific papers. Among the academic community I have heard the imprecation "publish or perish." It is generally used as a condemnation of the "system," indicating that one will not manage to progress in the academic community if he does not publish. This is both false and correct. Teaching a subject well does not necessarily imply that the teacher is at the cutting edge of the technology, nor does the converse work—the man at the cutting edge of the technology is not necessarily a good teacher! These are two distinct talents and should not be confused, one with another. A few gifted people have both talents, but that is not the general case!

For the would-be consultant, however, the requirement to be at the cutting edge of the technology is an imperative. The people who would be your clients are interested in your particular talent, at least in some cases. It is therefore important that you do some noteworthy work in your area. It follows as a corollary that work worthy of note is published; else how would your fellows benefit from it?

Now make no mistake about this matter either. The mediocre thesis published on an obscure topic will help not one whit in obtaining a private practice! The succinct paper that deals with real problems of the real world will. Think about it! I am sure there are a number of references that you regularly use in your work—the technical papers or texts which describe the processes you have adopted and found to be effective. Would you consider the author of such a reference to be an expert? I'm sure you would! The fact of the matter is that this person has helped you in technical matters in the past through his papers. If you have a problem, you would be pleased to meet him and would welcome the assistance.

There is something to be said about another aspect of the technical paper. Where do you air the subject? There are essentially three different places in which a technical paper may be presented: the trade magazines, the professional journals, and the symposia and meetings. Each has some distinct features and advantages.

The trade journals—that is, the free magazines sent to you monthly with the "Bingo" card—have far and away the widest readership. There are probably ten people who read or at least leaf through these magazines for every one who reads the *Transactions of the Professional Group on*——. The treatment of most of the subject matter is relatively informal, and the subject matter is practical. Furthermore, the articles are usually interesting. An author is prone to write about such matters as the implementation of a telephone dialer using the SC/MP microprocessor, and to include an anecdote about a programming error he made in the course of the research. These publications tend to contain short single-page write-ups on circuits that uniquely solve some practical problem in an optimum manner. Precisely because of this light but practical touch these publications are widely read. The overwhelming majority of the technologists in this world are not much given to wading through heavy theoretical treatises. Most of them, on the other hand, will read a well-written article on a practical subject in which they have some interest.

The professional journals usually are much more formal, and in general theoretical. A comparison of titles tells the story. A title from a trade journal is likely to read "Input/Output Interfaces Make the Digital Connection," while a title from a professional journal may be "Induced-Current Effects on Microwave Backscatter." The professional journal is much more likely to contain the results of studies of a general nature, performed at a research institute or an industrial laboratory, probably under government sponsorship. The readership of the professional journals is much smaller, and the going a great deal tougher than in a trade journal. Even among those who regularly read the professional journals, I estimate that the average reader reads no more than two or three articles per issue, and those are in his own specialty area. He will seldom do more than scan the titles of the remaining 20 or 25. The professional journals are the correct vehicle for reaching the fellow practitioners of your specialty with basic or fundamental work.

Symposia and meetings offer something more than the other publications do: you get to meet and see the audience. These symposia come in several flavors. Some—those associated with a trade show—tend to have the character of a trade magazine. The dominant theme is practical matters and practical solutions to engineering problems. Others, usually without a trade show, more nearly resemble the professional journals and present more basic knowledge. A very few are neither basically scientific nor bascially practical, but are largely political in nature. These latter are usually held to establish standards for telephones, radio frequency allocation, computer interface bus design and the like.

A paper presented at one of these meetings usually cannot be published elsewhere and may only appear in the printed minutes of the meeting. Thus you will, in most cases, not reach very far beyond the meeting itself for an audience. However, the symposium does have the advantage that the speakers get to meet and socialize with their fellows. It is important that one get to know the people who are principal contributors in his field, and the meetings are the easiest way to do this.

Each publication medium has a unique virtue, which should be utilized if possible. You will meet and get to know your fellow practitioners at the meetings. The professional journals will carry your basic studies in your art. And last but not least, the trade journals will reach the largest segment of your potential clientele.

WORK ASSIGNMENT

Another aspect of preparation is the work assignment. It is most important, since the experience acquired during the preparation period is the experience you will draw upon in dealing with your clients.

In most industrial organizations a few of the technical people always seem to be playing shortstop—for example, the fellow who is called in when the equipment does not function for the customer, when production suddenly breaks down because the equipment cannot be made to meet specifications, when the customer requires a change in design in order to solve his problem. This type of job is a very good background for the would-be consultant because it provides experience in two important areas: (1) problem solving in the most expeditious manner and (2) soothing the customer's feathers.

If you really want to be a consultant, this is the type of job you should work to get, even if it takes a fair amount of doing. Note that this kind of job implies that the practitioner is highly knowledgeable in the *whole* product and the end use of the product. You may be an expert in some subsection of the product; however, in order to perform this type of job, you must be fully competent in all aspects of the product and well versed in the things that the product may be used for. No matter how well you know the product itself, dealings with the customer require that you consider what *he* is trying to do with the product and understand *his* problems. He is interested in your product only to the extent that it will accomplish his ends.

WHEN AM I READY?

Presuming that the would-be consultant has gone through a number or all of these preparatory steps—he has written a number of papers, is reasonably well known in the trade and is regularly called out for problem solving—is there any criterion by which he can know that he is ready to have a go at consulting?

Probably like getting married, or starting a family, there is really no time when one *knows* that he is ready. There are a few guideposts, however, which will help. First of all, you know that your publications have reached a number of people when you start to get an occasional letter or phone call from outside the firm with regard to something you have written. With regard to problem solving, a reasonable criterion is the point when the annual total of your air-fares exceeds 25% of your gross salary. Financially, you should have some money in the bank—but that will be dealt with separately. Mostly, people start this sort of thing at the wrong time anyway, for there never is a right time.

At the very least, it should be obvious that the person who wants to try to establish a private practice immediately after graduation has more nerve than judgment. At this point in his career he has very little experience, and his circle of acquaintances in the profession is quite restricted. These factors translate into a substantial impediment to success. There are certain fields, such as social work or political science, where an outstanding academic record might provide the key to success; however, this is seldom the case in consulting engineering. The few engineers who have made a success of an independent career at this point in life—for example, Edwin Land and Stanley Hiller—have done so on the basis of outstanding inventions. Furthermore, their success was principally as entrepreneurs. They started a small business, which was their principal client.

At the other end of the spectrum, there is the relatively common desire to establish a consulting practice after retirement from a successful career. It is not unusual to find that a number of firms will retain the services of outstanding performers on a "consulting basis" after retirement. In most cases, however, this is simply a continuation of the pre-retirement job, arranged to get around the niceties of the mandatory retirement rule. The firm continues to have the services of the man on a reduced work load, and he has the benefit of useful work and added income with a bit more freedom to move about than he had as a regular employee. This can be a very pleasant experience for some people.

In the sense in which we are discussing a private engineering practice, the after-retirement start has a severe drawback with respect to time. As will be discussed later, it takes six to eight years to establish a satisfactory fulltime consulting practice. If the practitioner retired at age 65, the practice would only start to roll well when he was 71 to 73, and there would be a great many 60- to 80-hour weeks in between. Not all people have the physical stamina for that heavy a schedule at that age.

In addition, a great many of the retiree's associates would be retiring at the same time. Thus he would be forced to develop a new clientele.

Of course, it can be argued that the retiree is not put in the financial bind that a younger person would be, presuming that he has some substantial retirement income accrued during his years of employment. The occasional commission acquired through a modest investment in marketing might be just to his taste and most rewarding in human terms—enough work to keep his hand in, but not so much that he could not get in a bit of travel or a round of golf.

For most engineers, however, the best time to begin a full-time consulting practice is in mid-career. At this point he will have had the opportunity to acquire the experience and to establish the circle of associates necessary for success. Also, he will have the stamina required to succeed.

3

How Do You Obtain Clients?
(And Make Them Heros?)

The problem of obtaining clients is, of course, a central issue in the survival of any person in private practice. You must have clients in a continual flow, and preferably you must have a rather large number of them, if you are to succeed. You do not want to be too dependent upon any single client because one of the chief virtues of the consultant is that he is disposable. When the particular task is finished, the client will be finished with your services. It is to be hoped that you will hear from him again when another requirement arises, but you must be prepared to see assignments end.

In actual operation, most technical consultants are introduced to their clients in the same manner that doctors and lawyers are—through other clients. People who have been served well will usually not hesitate to recommend your services to others. When one moves into a new community and requires the services of a physician or an attorney, he asks a friend, acquaintance or neighbor for a recommendation. Except in an emergency, he will rarely just walk into an office. The matters to be dealt with are personally important, and he would very much like to know something about the person before he makes a commitment.

In many respects, the problem of a man requiring the services of a consultant is similar. He will have some problem requiring solution and will have been persuaded by someone or something that he should seek help. If he owns the business himself, he will likely see the money that the assistance will cost as coming out of his own pocket. Before he takes the step, he must be sure that it is the most effective thing to do. If, on the other hand, he is employed by someone else—for instance, a large corporation—he may have been directed to get in an expert. If he gets in the wrong person, the job may not be finished to the satisfaction of his superiors, and he will be considered guilty of poor judgment. The problem is usually compounded by the fact that, in either case, the man is not versed in the field of expertise of the consultant. If he were, there would probably be no consideration of outside help.

From the point of view of the client, he is faced with a situation that he cannot personally solve and is forced to take some chance with a person he does not know and whose exact qualifications he is unprepared to judge. Such are the joys of management!

Thus the typical first contact a consultant has with a new client begins with a phone call that goes something like this: "Mr. Kuecken, my name is . . . with the . . . Manufacturing Co. Joe Fotzengargle of Superior Flugel Pin tells me that last year you did some work for them. We have a similar problem . . . ." The fact that someone *he* knows was willing to recommend you has given him enough confidence to initiate the contact. This is a fairly basic rule in the consulting business. Only rarely do you encounter a client who actually wants to commit himself without some prior knowledge and recommendation. A few of the large consulting firms have a good enough reputation so that a *personal* recommendation is not required, but for most of us it is a necessity!

About now this sounds a bit like the argument of St. Thomas Aquinas. The blacksmith may continue to weld new links onto the hanging chain, but somewhere there must be a pin from which the whole thing hangs. Just where does one come by the first client?

Almost without variation, this first client and many of the first clients you will have come from the "clients" you had when you were working for the parent firm. It is not unusual to find that it takes two to three years in a consulting practice before one has a client who did *not* know him somewhere else. In order to survive, you will have to sell to people who knew you and whom you helped before you went into the business. This is one of the reasons why the problem-solving work assignment is such an important part of the preparation for going into the consulting business—it is your opportunity to get to know people and to demonstrate your abilities to solve their problems.

One feature of our technology works very kindly in favor of the would-be consultant. American technology is a remarkably fluid structure. The technologists themselves are a mobile group and do a good deal of swimming about. The stronger the person, the wider he may swim. After a period of 10 or 15 years, you find that many of the people with whom you have worked are diffused throughout the container. This is a distinct advantage to you in that it broadens the base of clients whom you will be depending upon to get started. The upwardly mobile manager will probably turn out to be your closest ally in your struggle.

There is an interesting corollary to this. Although there is probably no way to obtain hard factual supporting data, my observation has been that the firms with a very low turnover rate and stable personnel roster are the least likely to seek outside help. It may be that the general absence of new blood in the organization extends to the attitude toward outside help. On the other hand it may simply be that the stability in the roster means that everyone has settled into his job and that no outside help is ever required. In either case, when you encounter such an ultrastable organization, you would do well not to spend too much time or effort on the courtship. The chance of reward is too slim.

HARDWARE

Throughout the length and breadth of our fabled land there are probably not five consultants who make a full-time living at it without occasionally delivering some hardware, or perhaps doing some hands-on work on the hardware. As pointed out earlier, the primary work of the consultant will be of the Ingenious Solution type. In a few of the tasks you receive, the problem will be to correct a procedure or technique to free up production. Somewhat more often a study and series of recommendations may be all that is asked for, and you will be called upon to deliver nothing but a report. However, the far more frequent problem will be to come up with a new product or to improve an existing product. In such cases, the task is not finished until the product is demonstrated to work! In other words, you build or direct the building of something that will, at the very least, prove the feasibility of your Ingenious Solution.

At this point some disparity will arise, depending upon the nature of your work. If you are an aerodynamacist and you are working on the improvement of the wing of an airplane, it is not feasible for you to attempt to do this with your limited facilities. Conversely, if your task is the development of a small electronic instrument, it may be entirely feasible for you to breadboard the design so that you have a workable (but perhaps not packaged) unit to demonstrate the operation. In my opinion, the consulting engineer should have the facilities available to at least try out his good ideas so that they can be shown to run. It is remarkable just how often they can be shown *not* to run! At this point a trip to the old drawing board is in order! If you have the facilities to breadboard things and prove the ideas, you will save a great deal of embarrassment for yourself and your client. Another facet of this situation is the fact that it is often very difficult, if not impossible, to decide which of a series of competing solutions will be the most practical and the most effective on the basis of theoretical study alone. Only when the hardware is tried will the answer be revealed.

To give a practical example, a number of years ago, I was approached by a client who had the problem of determining the level of a black powdery substance within a polyethylene container, which was used in a machine that the client manufactured. If the powder ran out, the machine would be damaged; therefore, the client wanted a reliable means of determining when the powder was low and interlocking the machine so that it would shut off and flash a "low powder" message.

The powder itself was about the consistency of face powder and had many of the physical properties of lamp-black. The most careful of mortals could handle it for a few seconds and look like a chimney sweep. It could pack down to a fraction of the "loose" volume and could easily bridge the outlet of the hopper, so they had installed a "thumper" to beat on the sides of the container to prevent bridging. This action blackened the sides of the container so that a photocell arrangement (which had been the client's original approach) would not work. As an approach to this problem I laid out a list of the electrical, mechanical and other physical properties of the material and tried to come up with a solution based upon each of these.

Since the material required agitation anyway, I considered using an electrically driven thumper and measuring the resistance to agitation. This technique worked; however, it turned out that the polyethylene containers varied enough from container to container, and indeed from place to place on the same container, that an unambiguous reading could not be obtained. It did not appear practical to maintain the level of uniformity in the containers which would be required so that this scheme would work unambiguously without a tailoring adjustment.

When the material was agitated, it had a tendency to flow around like a liquid. It seemed that a float mechanism would easily determine the level. Unfortunately, when the hopper ran low, if more powder was dumped on top of the float, it took an inordinate time to rise to the surface, even with the thumper running. This could be prevented by having the fillee hold up the float with one hand and dump powder with the other, but that left the fillee pretty filthy.

As an alternative fluid approach, I placed a small tube near the bottom of the hopper. A small, low-volume vibrator air pump, such as those used to pump air in an aquarium, was connected to this tube and the back pressure measured with a small diaphragm and switch arrangement. When the powder was above the end of the tube, the back pressure held the switch closed. When the end of the tube was uncovered, the pressure fell and the switch opened. The pump was not capable of enough volume output to seriously blow a lot of the dirt and powder around, but the pump had enough pressure to blow a plug out of the end of the tube if the material packed-in during filling. This was actually a pretty workable scheme, but it did require the services of the air pump and required a pressure-tight connection to the hopper.

Another scheme was based upon the electrical properties of the powder. It turned out that the powder was a dielectric; that is, it did not conduct electricity. It had a dielectric constant of 2.73 in the loose form—which is to say that a capacitor filled with the material had 2.73 times as much capacitance as the same capacitor filled with air would have. The basic electrical scheme was to measure the capacitance between a center bar and two outside bars. If the gap was filled with powder, the capacitance would be 2.73 times as great as it would if the powder was gone and the space filled only with air. The circuitry for this arrangement was simple and cheap. It required only three transistors and an SCR to provide the required switching logic.

This electrical scheme worked quite well. The principal difficulty to be overcome was the fact that when the material packed or bridged, the dielectric constant rose. Therefore, a packed condition in the electrode area with portions of the area open could potentially give a proper indication while no powder was actually flowing. This condition was postulated and forcibly induced, but it did not ever occur during the experiments.

A mechanical scheme was also investigated. A small motor-driven paddle was used to stir the powder. When there was no powder, the load on the motor dropped off and the motor drew less current if a DC motor was used. With a cheaper shaded-pole AC motor (the type used to run a phonograph turntable, etc.) the change in current was too small to fire on reliably; however, the phase angle of the current changed significantly so that reliable firing could be obtained on the no-powder condition. For the AC device, the motor was cheaper but the sensing circuitry was more expensive. If the motor simply wiggled the paddle through a flexible part of the container wall, the problem of false firings due to dirt in the bearings was essentially eliminated.

Of the workable schemes, the air-operated back-pressure scheme required the least modification to the hopper. The dielectric sensing scheme used the fewest and cheapest components but required modification of the hopper to include the electrodes. The motor-driven scheme required some modification to the hopper but was probably the most expensive to implement. It probably could have been used to fulfill both the functions of the thumper and the sensing, thereby eliminating the cost of the thumper, which would have made it more attractive economically.

Now the point of this example is to indicate the requirement for experiment. One could have spent an endless amount of time debating the merits of the various schemes, and a committee would not have reached a conclusion to this day regarding which of the approaches would be the best to investigate. The actual time involved was just a little over a week including the typing of a slightly smudged 21-page report. (Even after careful hand washing, the powder would gradually bleed out of the pores!) While the experiments did not settle everything, they eliminated a number of unfruitful approaches and gave a realistic "feel" for what was involved and the actual costs and problems of the finished item.

This is a point to be stressed rather strongly, since the viability of a consulting practice can hinge rather heavily upon it. The small consulting firm can nearly always be cost-effective in simple gadget situations such as this. You have an idea and you can rig something together in an afternoon and try it. Your technician can take a free-hand sketch from a lined pad and produce the pieces required in a few hours or a day. By comparison, the corporation engineer would have had to make up fairly precise and detailed sketches for the model shop and would have had to obtain a requisition for the model shop services. The model shop, in turn, would have had to requisition material and assign the task to an individual model-maker. Depending upon the formality of the procedures of the client firm, the effort could have been extended considerably more.

Make no mistake about it—the simple experiments described did not in any sense represent a finished product, nor could the client have gone straight into production on any of them without courting disaster. He requires a careful specification, design and documentation procedure in order to turn out a good product in some orderly fashion. His organization is optimized for that type of operation, and yours is not. One of the chief benefits of small size is agility—which is where *you* should be optimized! The ability to respond quickly and effectively to situations requiring a number of small experiments can represent one of the most salable aspects of your consulting services.

INVENTION ON COMMAND

In connection with the ability to respond, the question is often raised as to whether a consulting firm (or anyone else, for that matter) can invent on command. Within a few constraints, which I shall outline, my answer to this question is always an emphatic YES! Furthermore, I submit that such invention is regularly done within many firms.

The first of the constraints that we shall consider is the question of Subjective and Objective Invention. Now, Objective Invention is a legal or social judgment made by the U.S. Patent office (or other patent office) or by your fellow workers in the field. Since it implies an action on the part of someone else, it obviously cannot always be done on command by the individual. On the other hand Subjective Invention is the unique and novel work of an individual (which may or may not be later adjudged to be an Objective Invention). It can nearly always be done on command by a perceptive and well-informed individual. It simply implies that when confronted with a real-world problem or requirement, the capable person will be able to come up with one or more reasonable solutions.

A second set of constraints is posed along with or as a part of the problem. Very often these are practical constraints that dominate the entire problem definition. A solution may be a solution only within the framework of the constraints.

As a direct example of this second type of constraint let us examine the critical natural gas shortage that developed during the extremely cold winter of 1977. The cry was frequently heard that there ought to be something that technology could do to produce a substitute for natural gas. And indeed there is!

When I was a boy learning to fly, the Detroit city airport was dominated by a huge tank that stood within the crook of the L-shaped airport. The tank was 385 feet tall and probably 200 feet in diameter (it was more important for a pilot to know the exact height than it was to know the diameter!). This tank was filled not with natural gas but with coal gas, which was cooked out of coal in the production of coke for the steel industry. People got along fine without natural gas.

Now we haven't stopped making steel and some of this gas must still be made. Why couldn't we use that instead of natural gas? The answer lies in the fact that this is a relatively low-energy gas. Roughly speaking, it has less than half of the energy content that pipeline natural gas has. This means several things. First of all, your furnace would not work very well with it, and, second, you would have to use at least twice as much. In most places this means that the pipelines are far too small to handle the amount of gas required. We would have to dig up the pipeline system and replace it in order to provide the necessary capacity.

Thus, one of the constraints on the natural gas problem is that the question is a little more sophisticated than it first appears to be. It is not just "Can't we find a substitute?" but rather "Can't we find a substitute with the same amount of energy per cubic foot so that the distribution system can be used as is?" As a matter of fact, this is just what the HYGAS and LURGI pilot plants are all about. These processes are aimed at producing a pipeline-quality gas from coal. The federal government and a number of industrial research foundations have been investing millions in this investigation.

The plants do in fact make pipeline-quality gas; however, the HYGAS program has until recently required hydrogen for the production of the gas. It is obviously a bit self-defeating to use hydrogen manufactured from petroleum in order to produce coal-derived pipeline gas. The hydrogen can also be produced by electrolysis of water, but this method is a bit expensive too. In the most recently developed system the hydrogen is derived by oxidizing hot iron in the presence of steam. The iron would be reduced and recycled in a subsequent process. Thus the complexity of the question grows to include economics, the use of scarce resources and the question of the air pollution produced by the burning of the coal.

A solution that does not fall within the constraints may be no solution at all. However, the point is that the people who have been involved in this investigation have been continually and step-by-step inventing solutions in the development. Some of these solutions may prove to be legally protectable Objective Inventions, while the majority will be Subjective Inventions that merely do the job.

HERO MAKING

Typically, the consulting engineer will not become involved in anything quite as grand as the development of a whole new industry to solve one phase of the energy problem. However, the constraints on the problem solution will be no less real for you and your clients. First and foremost is the economic constraint, in most cases. There is usually some fixed price above which a piece of consumer or commercial goods will not sell. A color TV would move for $400 in 1977, but you would have had a great deal of trouble selling a set for $700 no matter how much better it was or how many bells and whistles it had. If you are going to make a hero out of your client, you had darn well better be concerned with learning the constraints of the problem and producing a solution within those constraints.

Likewise, it is well to remember that the cost of development has to be amortized over the sale of the product. If your development cost were $70,000 and it were spread over the production of 7,000,000 cars, it would work out to a penny a car. On the other hand, a development cost of $7,000 performed on an antenna coupler expected to sell 1,000 units in the first two years would work out to $7 per coupler. This is a highly significant consideration because most of the items a consultant gets to work on are relatively short-run items where the development contributes significantly to the sale price.

Another significant constraint stems from the capability of the client to produce. If your solution to a problem requires work with miniature precision bearings and gears, it might be suitable for a watchmaking firm but not for a radio manufacturer. For the latter, something with a board full of integrated circuits might be just great, but anything with moving parts might prove to be a disaster. It is a fairly good rule that a mediocre solution which departs little from what the client is doing now is often better received and more successful than an elegant solution which requires a complete change in methods. This is a bit of a hard lesson to learn, but in problem solving always be sure to consider and present some minimal solution. The elegant solution should be there too, but it may be too expensive or risky for the client to try.

There is a tale told about a physics professor who asked his class how they would find the height of the Memorial Tower if they were given a precision barometer and a yardstick. Most responded with the standard solution, namely, to measure the pressure of the atmosphere at the base and the top and calculate the height of the tower from the difference, assuming that the pressure falls one inch of Hg per 1,000 feet. The second most common answer involved measuring the length of the shadow of the yardstick and the length of the shadow of the tower. A few had more unusual solutions. One suggested dropping the barometer from the top and timing the fall, whereupon the height could be calculated by;

$$S = \frac{32.2t^2}{2}$$

where t is in seconds and S is in feet. This works but is hard on the barometer.

One student suggested lowering the barometer to the ground with a wire and then using the yardstick to measure the wire. The final one suggested going to the plant engineer's office and offering to give him the barometer in return for a look at the building plans, where the tower height could be read. In all likelihood the last two solutions would yield the most accurate result and would be easier to perform than the others. Now if the problem had been limited to finding the height of the tower, the last two would have been the best solutions. Since the real constraint was to pass a physics test, these answers were unacceptable! ALWAYS MAKE

SURE THAT YOU COME UP WITH AN ANSWER THAT THE CLIENT IS READY AND ABLE TO TRY!

You will note that I have been emphasizing multiple solutions to problems, not just one good idea but a number of them. It is rare for a problem not to have a number of different solutions. They are not all equally good, nor, as we have seen, are they all equally workable. However, you owe it to yourself and to the client to consider the best list that you can conjure up if he wants you to solve the problem. Then, if at all possible, try them to see how well they work and what other problems the solutions may raise. You may want to present all of the workable ones; however, be sure to present the simplest and most workable FIRST! Elegance and rigor have a tough time winning over the simple and straightforward.

Another item to keep in mind is that the single most common complaint about the use of a consultant is that the first thing he will recommend is the requirement for more consulting service. Nothing else will turn off a client faster. Except for the little one- or two-day "come and look and give us a recommendation" type of task, you will find that without variation you will be working on fixed price or not-to-exceed purchase orders. The client will ask you to quote him a price and may occasionally fund that effort; but, for heaven's sake, try to do the most thorough job possible within that price. THE VERY BEST WAY TO OBTAIN MORE CLIENTS IS TO HAVE THE ONES YOU WORK FOR FEEL THAT THEY GOT A BARGAIN.

The next directive concerns the subject of travel expense. Most consultants charge travel expenses if out-of-town travel is required. It is very good policy to send a travel voucher similar to the one you used when you were working for the corporation. You can buy these with a self-duplicating carbon at most office stationers. And when you travel, keep the expense within reason. If the client has his employees fly tourist class, you do too—or pay for the difference out of pocket. The chances are good that the manager who retained you will receive the voucher from Purchasing or Accounts Payable, and it will do you no good to have included expenses that he would not have been allowed!

Another item, which comes under the heading of common courtesy, is simply: <u>NEVER SHOW UP A CLIENT'S PEOPLE</u>. When you are brought into a strange plant and introduced to a whole new crew of people, it is common for at least one or two to be a bit suspicious or apprehensive about your presence. Mostly, they will be cordial, and a few may be a bit awed by your presence because they know that management went to some pain and expense to bring you in. Sometimes a few will feel that they have been challenged, and they may feel that the whole effort is a farce. Such a person may just be the guy who is irascible to everyone anyway. Most firms have one or two of those. The point is that you are a *guest* in their province, and you should always remember to behave like one! If it is humanly possible, try to get along with everyone and <u>NEVER BAD-MOUTH ANYONE.</u> You will have only a fleeting glance at the organization, and that should be concentrated on accomplishing the task for which you were retained. The manager and the other workers are probably all well aware of their other problems, including personnel. A careful person can usually point out what needs to be done to solve an assigned task without specifically pointing a finger at anyone in the organization or offending anyone.

HUMAN CONSTRAINTS

As noted earlier, problem solving can be constrained technologically and economically. However, the largest and most significant constraints ever to be faced by a problem solver are and will continue to be human constraints.

Consider, for example, the plight of the British/French consortium that built the Concorde:

The Common Market nations are faced with the problem that the market for jet transport aircraft is dominated by U.S. aircraft builders. Even Air France and BOAC, which are partially supported by the respective governments, are mostly populated with U.S.-built jet liners. At the same time that their aircraft industries are languishing, their own governments are forced to purchase U.S.-built aircraft for economic reasons.

The solution proposed by Britain and France was to form a consortium and, with government subsidy, to fund the development of a technologically superior aircraft. If the planes were superior enough, the various lines would be forced to purchase them, even at a premium, because the passengers would not be content to ride the slower planes. Who would want to spend six hours crossing the Atlantic when it could be done in three?

The project was given an early boost by the cancellation of development of the U.S.-built SST. It became apparent that they would have competition only from the Russians for their supersonic airliner.

At the outset, the Concorde was plagued with what seemed to be overpowering political and social problems. The airplane could easily cross the Atlantic in three hours, but, when it got to the American side, there were a great many people who violently objected to having it land. The people living in the vicinity of Kennedy International Airport in New York complained that it was even noisier than the regular jetliners that were already assaulting their ears every few moments. The social reaction was so strong that groups became belligerent and deliberately blocked the approaches to JFK with their automobiles in a slow-moving procession.

The United States Congress also got into the act, with several of the representatives delivering impassioned speeches regarding the noise pollution due to the Concorde's very large and noisy engines and the air pollution due to the massive fuel consumption it required in slow flight. It is always difficult to separate chauvinism from honest concern in such matters, but it is interesting to speculate whether these members of Congress would have found the Concorde to be quite so noisy if it had been built in Seattle, Long Beach, California, or St. Louis.

At a certain point, the Concorde must have seemed to the British/French consortium to be a technical triumph and a financial disaster. For, without permission to operate the airplane on the New York run, there would be no way that the consortium members could sell enough airplanes to recover their investment. *The Concorde was proving to be no solution whatsoever to the main problem—that of revitalizing the British and French aircraft industries.*

This problem has been overcome—mainly in the political arena—and the Concorde is regularly landing in New York. The furor about the noise seems to have faded from sight since the battle over landing rights was politically lost. The British/French consortium will not have the answer regarding whether the craft will revitalize their aircraft industries for another few years. The aircraft seems to have a very high cost per passenger-seat mile, and it is very small compared to current jetliners; so, the craft may yet fail in the financial arena.

The consulting engineer will seldom find himself confronted with human difficulties on such a grand scale; however, the "political" constraints on problem solving extend down to the smallest transactions. At the corporate level such cases are not entirely unknown and must occasionally be dealt with. On occasion, one finds that the solution to the problem is actually at hand, but its application is being prevented by simple human jealousy or a power struggle within an organization.

Consider the case of a firm which manufactures a particular product that requires great precision in the fabrication of its parts. Initially, when the product was introduced, the machines were newer, the firm had very skillful operators, and the required standards for the operation of the finished product were modest. With passage of time, the highly skilled operators have retired, the machines have become worn, and the press of competition has raised the performance standard for the product. An acceptable yield is no longer being achieved.

The Manager of Manufacturing maintains that if he were allowed to expand his department with automated digital-controlled machinery, he could have the situation in hand within six months and manufacture suitable parts at less cost. At the same time the Manager of Engineering maintains that if he were allowed to expand his department by hiring some electronics people, he could eliminate all the precise mechanical parts from the product. Thereafter, the manufacturing operation could be substantially reduced while the same product output would be maintained, so that cost would be reduced.

Marketing sides with Engineering because they feel that the electronics will add sex appeal. On the other hand, Field Service questions whether it can adjust to the new product and sides with Manufacturing. The Comptroller sides with manufacturing as well, since he foresees risks and costs in hiring new personnel and unpredicted engineering difficulties.

The consultant called in by Corporate Management to report on the situation treads a very fine line indeed. Actually, both proposed solutions are probably workable, and the difficulty is principally one between the personalities involved. Unless there is some very strong evidence in one direction or the other, the consultant is liable to come out of the situation with a black eye.

I would not suggest that these problems are the common case or that the example is not drawn with a heavy pencil, but some of this kind of thing is involved in perhaps 10% of the cases encountered. The people involved are sensitive and do have pride. As an outsider, it is incumbent upon the consultant to try to fulfill the assignment as well as possible. IF IT CAN BE AVOIDED, NEVER PROVE ANYONE IRREVOCABLY WRONG!

SUMMARY

Taken all in all, the task of making a hero of the client boils down to:

> DO A GOOD JOB.
> DO IT FAST.
> DO IT AT MINIMUM COST.
> DO IT POLITELY.

The client thus treated will bring you the next client!

4

Fees And Contracts

The matter of establishing fees for your services is most important in any kind of practice. Nearly everyone would like to optimize his income, but doing so does not necessarily correspond to charging the maximum rates for one's services. There is obviously a point where people will simply bypass you because they cannot afford to have your assistance. On the other hand, there is obviously some saturation point at which you can no longer increase the volume of low-profit work that you can accomplish. The optimum fee rate must lie somewhere in between.

One guidepost in this wilderness is the rates that are being charged by the college professors in your area for consulting services. To most of these people, this income is not a major part of their livelihood; therefore, their rates may not be optimized. However, such data can serve as a useful guideline. But it is important that you be scrupulously honest with yourself in the comparison. Try to find someone who has about the same skill and experience level for the comparison. There is, after all, some cash value to the prestige of being a department head or a Nobel laureate.

Another worthy guideline is the salary level of the people in industry with roughly comparable skills and achievement. For the typical large corporation the overhead rate applied to engineers can run between 108% and 200%, with a figure on the order of 125% being relatively common. This overhead goes to supply the roof, heat, light, telephone, secretarial service, employer's contribution to the health plan, the retirement fund, Social Security tax, and so on. You are going to have to supply each of these expenses out of pocket; therefore, it seems reasonable that they should form a portion of your overhead as well.

On this basis, let us suppose that the comparable engineer had a salary of $25,000/year. If we multiply this by 2.25 for an average overhead rate, we arrive at $56,250/year. Assuming that a work year amounts to 250 days, we arrive at a rate of $225/day. This is what that engineer is actually costing the company, and a savvy engineering manager will know this number to the penny. To be cost-effective, the work you do for the company must save enough days of his effort to carry the bill.

Of course, there are cases where the costs to the company are astronomical, since some entire procedure is shut down or held up. During the North Sea oil well blowout, the losses on a per day basis were enormous. The money paid to "Red" Adair and his group was peanuts by comparison. Probably very few things that one can do are as profitable on a per day basis as putting out oil well fires and capping blowouts. While this is an extreme example, similar situations follow on a smaller scale. For example, an oil tanker that cannot be delivered to the customer because the antenna system will not perform to SOLUS specifications can cost $85,000 to $100,000 per day sitting in the dock. Depending upon the nature of your services, the "money-is-no-object" assignments could range anywhere from 0 to 20% of your efforts. However, for the most part, you should plan to arrange your fee structure so that it is cost-effective to the people who use it.

It should be borne in mind that even "Red" Adair does not have a blowout to cap every day, and that a certain fraction of your time will not be billable to anyone. Consider as an example the new-product situation:

A prospective client calls on the telephone and says he was told that you developed a new product for someone he knows. His firm is considering a similar approach, and he wonders whether you would be interested in dropping in to discuss the matter. On the first call you visit his plant and spend some time meeting his people and discussing the nature and requirements of the new product. Presuming that the effort is within the range of your capabilities and that you have something to suggest that catches his fancy, you will be asked to prepare a brief proposal. He would like to have a letter describing just what it is that you would propose to do, how long it would take, just what it is that you will deliver and how much it will cost.

This is a pretty reasonable request. The chances are good that the man you dealt with does not have the authority to commit his firm to some relatively expensive enterprise on his own say-so alone. He will have to have a written statement describing the development in some detail in order to obtain the approval of his superiors for the commitment. Accounting will have to have a statement of costs, since no one there has the authority simply to sign a blank check and hand it to you. Marketing will have to have a description of the product you expect to be able to produce before they can reasonably judge whether they will be able to sell it. Furthermore, they will have to know what the finished item is going to cost and when they can have it off the production line. This brings in Manufacturing and Engineering.

If you are reasonably quick-witted and agile, and you do not have any great delays in finding the costs to you of the material to be supplied, you might be able to have a small proposal letter out in a week or so with the expenditure of perhaps two days of your efforts and some long distance calls to suppliers. You now have three days of consulting effort invested in the project.

Now on the other end of the line you will note that there were a great number of people involved, representing several different disciplines. Each of these people is going to be asked to render a professional opinion on the subject. Engineering is going to have to make up parts lists and give some description to Manufacturing in order to obtain some reasonable estimate of the cost to build the final product, Engineering will have to provide an estimate of the engineering needed to reduce your breadboard to a manufacturable device and prepare the necessary parts lists, drawings and procedures. Marketing will have to somehow survey the market, after the pricing exercise is finished, to determine whether they can sell the finished product.

These matters take time. Unless the product is a very small and simple item, you can expect to wait from one to six months to hear the verdict. In the meantime, you will have to expect to make telephone calls and answer questions and perhaps keep a few more appointments before you find out whether the thing will fly or not.

Even if you are quite successful, the capture ratio on things of this sort is less than unity. You can always expect that some of the things you have put effort into proposing will never materialize. It is therefore important that you have more proposals out than you could fulfill, so that the ones that do come through will be adequate to sustain you.

The long time constant on these things is the principal reason why it takes a long time to get a consulting practice started. It simply takes a while to fill the pipeline with marketing effort and proposals so that they start to flow in with any regularity. At the outset you will obtain only the short-time-constant ones, and the work flow will be sporadic. After a few years have gone by, repeat business begins to fill in the rough places, and the flow will smooth out. It is not that the long-time-constant first efforts become any less so, it is simply that the repeat business on the shorter-time-constant efforts tends to spread out to fill things in.

The short-time-constant programs are usually the problem-solving and study efforts. In problem-solving situations the time constant can be *very* short, like, "When can you get the next plane?" In the case of study efforts, the time constant is more likely to be on the order of a week. Here the client usually will have decided that it is advisable to call in a consultant to look at the situation and write a report. He will probably ask you how long it would take and what your rates are. He will have a preconceived notion of the correct numbers. Chances are good that the money for the effort will have been tentatively allocated before his call.

An important subclass of the study effort is proposal preparation. The client is preparing a proposal of his own for a potential customer and feels that he would like to beef up his proposal effort with the services of a consultant, who presumably has some expertise and reputation in the field.

Proposal efforts are always hectic and usually fun. The effort usually assembles a group of the most imaginative and productive people that the company can scrape up and lumps them together to interact. Particularly in the case of military proposals, the specifications are relatively iron-clad and often self-conflicting: "Not less than . . . watts" and "Not more than . . . pounds," or "Not less than . . . microvolts per meter at 1 mile," coupled with "Antenna height not to exceed . . . feet." Proposal time is a time for agonizing over which of the requirements the customer might "give" on in order to satisfy the other requirement. It becomes a sort of mental chess game with attempts to devise a winning strategy.

An important note must be made on the subject of proposals. For many consulting engineers, proposal preparation is an important part of the practice. In any given year it can account for a major portion of your income and will generally run in excess of 25%, especially during the first four or five years. It is necessary to protect this income if you are to succeed.

When you are starting out, it is quite common for some small firm to call and announce that they are preparing a proposal and would like your assistance. This is a very good time to announce your rate and billing structure. You will then sometimes hear, "Well, we had thought that you might like to participate with us in this effort. Of course, if we get the contract, we'll be happy to let you bid on doing the work." There will then follow a large number of assurances that they have a real inside track on the effort and your return is nearly assured.

There is an old saying among the practitioners of the Oldest Profession—"Never give away what you can sell."

In my humble opinion, this sort of offer should *never* be accepted for the following reasons:

1. It is not fair to other clients whom you will want to charge full list price.
2. You have picked a place to put your venture capital, your own business. Do not allow someone else to place you in a twice-removed risk position.

Actually, the chances of success on a deal of this sort are usually vanishingly small. I know several businessmen who have made a practice of obtaining gambled help at any time that they felt that they had little chance of success but wanted to "show the flag" in order to stay on the bidders list. When a job comes along that they think they have a chance of winning, they will spend the money to bid it! On the rare occasion when they do succeed on one of these "show the flag" efforts, the people who prepared the proposal for nothing will seldom even be contacted.

At the beginning of your practice when there may not be much else to do, it is a bit hard to turn these people down. I feel that you would be well advised to do so. In order to succeed, you will need the respect of your clients and the opinion that you are working in professional earnest. Being duped into "freebie" deals is no way to enhance your position!

Before closing the discussion of fees, I would submit that ex-cessively low fees or backing off from the fee structure will not enhance your position. First of all, if you announce a fee struc-ture that is lower than usual, the clients will likely interpret it as representating less value. Also it is distinctly unprofessional to dicker about rates. If you have managed to establish a reasonable rate, it is advisable to stick with it. I cannot recall a situation where I felt I lost a task that would have been gained had I asked for a 25% lower rate. Clients expect to be paying a premium for a consultant, and they are either ready to do so or not. If they are not, they would not have made the deal on a 25% lower rate structure.

There are, of course, certain situations in which the client simply does not have quite enough money to cover your proposal. This occurs most frequently in new product development and special instrument development situations. In such cases, if you still want to do the job, it is usually more graceful to back off by suggesting that you do less, even if you know that you may have to do al-most exactly as much. If you are clever, you can usually find some place where you can trim what you are offering slightly without losing major benefit to the client. In connection with this, there should be major attention to the contract as discussed in the following section of this chapter.

To summarize: ESTABLISH YOUR FEE STRUCTURE TO BE

COST-EFFECTIVE TO THE CLIENT. DO NOT LET IT GET TOO FAR OUT OF LINE WITH YOUR FELLOWS. NEVER HAGGLE!

CONTRACTS

The material that follows is in no sense to be construed as legal advice. You can get that from your attorney. What is intended is to provide a few pointers based upon some years of experience, with the hope that you can avoid some of the pitfalls.

Patent Rights and Confidentiality

Patent rights and confidentiality constitute one of the most important aspects of dealing with a client. In probably no other area can bad feelings arise as easily as in these areas. The client can be very touchy about the subject of patent rights and confidentiality.

To begin with, the client who brings in a consultant frequently does so on the basis that the consultant may contribute some new idea or technique. This new material may in fact prove to be Objective Invention. It may also be something that can improve the competitive position of the client. The attitude of the client usually is that he has bought and paid for the work which led to the generation and reduction-to-practice of these ideas. Hence, he will probably feel that he is entitled to the free use and exclusive possession of such material. To this end, the client may insist, before the issuance of a purchase order, that the consultant freely assign all patent rights to any new or novel material generated in the performance of the commission.

Similarly, the client knows that the consultant will have access to certain material that the client has developed at his own expense and which he considers to be "proprietary," i.e., his sole and private property. Since he knows that the consultant is an independent agent who might well do business with a competitor at some later date, the client may include a confidentiality agreement along with the patent agreement. His legal department may insist on the execution of these documents prior to the assignment of a purchase order.

Among engineers with whom I deal, it frequently comes as a

surprise that I will freely assign the patent rights for inventions developed during the course of a funded study. To many of them one of the chief attractions of being self-employed is the retention of patent rights, with the presumption that they could manage to profit from the royalty income accruing to the licensing of these rights. The vision of having this form of annuity to support themselves during their retirement years is one of their strongest inducements to break the shackles that bind them to the corporation. A good many of the more productive engineers have seen one of their bright ideas become the backbone of the business of the corporation and the source of many millions of dollars in income. In return, they themselves have received an invention award of $100 and favorable treatment at their next salary review. To many, this seems unfair.

In contrast, very few feel that the proprietary rights clause is unfair. Most feel that the corporation has property rights to techniques and designs developed by the corporation with its own funds. Most would agree that a consultant who would peddle a trade secret to a competitor would be stealing.

At the personal level, however, proprietary rights receive a somewhat more muddy interpretation. Suppose that engineer A has a job with the XYZ Corporation. In the course of his work he is called upon, as a job assignment, to develop a new type of flugel pin. It is immensely successful and brings the corporation great profit. He receives the invention award and an increase in salary. However, after some period of time he feels that his recognition is not commensurate with the large contribution he has made to the XYZ Corporation and he begins to entertain offers of employment elsewhere.

In such situations the engineer will solicit offers within the industry that has employed him. That is where his experience lies and where he could expect to obtain optimum benefits from it. As a result, he is most likely to obtain advanced employment from a competitor, such as ABC Flugel Pin. No one really expects to have him arrive at his new job as an amnesiac with no recollection of any of the work done at XYZ; as a matter of fact, if he did receive an advancement from ABC, it was precisely *because* of his previous experience. The question then becomes one of whether what he does at ABC represents a disclosure of unique proprietary property of XYZ or instead is merely the application of personal experience to a new job. The experience is usually felt to be the unique property of the individual.

Obviously, this is a situation with shadings of gray rather than a binary black and white. The frequency of lawsuits in this particular situation is not large and probably involves only a tiny fraction of a percent of all job transfers. However, the courts have been known to enjoin Mr. A from working on flugel pin designs relating to his previous assignment. In other cases, they have found that XYZ has no right to prevent Mr. A from obtaining employment making use of his experience.

In the case of the consultant, my feeling is that the client has every right to expect that his property and position in the market will be protected to the maximum degree possible by the consultant. This extends to the case where the consultant may have to turn down a commission because it is too directly tied to areas where previous work for competitors has placed significant amounts of proprietary information at his disposal. This type of conflict-of-interest decision must be made by the consultant alone. It is well to remember, however, that the suspicion by others that it was not made honestly can cost dearly in terms of future business.

With regard to patent rights, I feel that these, too, are proprietary rights of the client. When the client retains your services, he is more or less expecting that your unique expertise will provide new and novel material, and he feels that it should be a part of the bargain. If you were to contract with a dealer for a new car, you would certainly expect to have the right to drive that car without the payment of additional rents, fees or royalties. I suggest that the same principle applies to patentable techniques and devices. If you finance the services of a consultant to develop something new, the result of his effort is yours and not his.

In the case of inventions, a few other points must be made. It is not unusual for a client to approach you with a commission for an effort in which you have made some considerable investment of your own. After all, it is likely that you would be working on new and novel concepts within the field of your expertise. The invention that you have independently developed and demonstrated may provide an optimum solution to his problem. Presuming that this effort is sufficiently well documented, you have two alternatives:

1. You may offer to sell the rights *prior* to establishment of a commission for the effort.
2. You may elect to throw the effort and investment to date into the bargain arrived at for the commission.

In either case it is important that the disclosure of proprietary rights be made *before* any agreement for consulting effort is reached. It would be very poor policy to accept the commission and then announce that you had a conflict of interest with the client owing to a previous investment on your part.

In many cases, if your investment is not too great or your position is not too strong, the most profitable thing may be simply to reveal your position in the proposal for the work and offer to throw your rights into the commission, with due regard for your interest in the pricing. It is usually much easier to do this than to arrange for the sale of patent rights.

Another consideration, which is frequently glossed over, is the fact that the preparation of a patent application usually takes a good deal more time and effort than does the simple signing of the documents. When properly done, a patent application represents a sequence of exchanges between the inventor and the patent attorney. The inventor would probably have to prepare drawings, write text and claims in model form and spend a certain amount of time in conversation with the attorney attempting to convey his concept of what the invention is, where the novelty lies and what it is important to claim. The attorney would then institute a search, and the results should be reviewed by the inventor. If the chances of a patent still looked good, the attorney would prepare a draft of the application, which should also be reviewed by the inventor. Only then would the signing come.

Now, the decision to file a patent application is usually something that takes time to reach. Most corporations have a Patent

Review Committee, which considers the various dockets submitted. Because of the cost of preparing a patent application, not all of these dockets are approved for filing. The results of the Review Committee are generally reviewed by several of the management heirarchy. All in all, it may take from four to nine months to obtain a decision regarding filing on any given docket. For the employee docket this delay poses no significant difficulty; however, in the case of a consultant docket, the fact is that the commission is probably over by the time that the decision to file is made. Under these circumstances there may be no mechanism for compensating you for the time required for the preparation of the patent application. This problem can be complicated by the fact that the Patent Department may be located in an entirely different part of the country, and be a portion of an entirely different division from the original client. It is therefore advisable to note in the original proposal that you are to be compensated at a rate of $——/day for time requested by the client for the preparation of patent disclosures and applications. A note covering travel expense necessitated by client-generated travel requests is also in order.

Summary of Patent and Proprietary Rights

It is important that an appropriate arrangement regarding patent and proprietary rights be arrived at prior to establishment of a working arrangement with a client. UNDER ALL CIRCUMSTANCES, THE PROPRIETARY RIGHTS AND COMPETETIVE POSITION OF THE CLIENT SHOULD BE PROTECTED.

Delivery Guarantee

In dealings between human beings it is most important that both parties understand what is expected of the transaction, just what is to be paid and just what is to be delivered and when. This can be a very great source of annoyance and hurt feelings.

If you have bought many automobiles, you have almost certainly experienced either a "highball" or a "lowball" at one time or another. If the salesman senses that you are shopping around, he is likely to try one or the other tactic on you. In the "highball" he offers you an unrealistically high figure for your used car. This has a tendency to shut out the other dealers and bring you back. In the final analysis, of course, the manager refuses to go along with the figure. However, you did come back, and he has a better shot at the sale. The figure the manager will agree to is probably competitive with what the other dealers offered, and you are tired of shopping by now.

In the "lowball" the salesman quotes you the base price of the car and "forgets" to note the fact that you asked for air-conditioning, power steering, and so on. The price quoted offhand is low enough to bring you back, and, of course, when the figures are added up in earnest the "forgotten" items reappear in detail.

In both of these cases, a swindle or, at the very least, a shady business practice, is perpetrated by the deliberate exchange of incomplete or misleading information about the transaction.

A CLIENT WHO FEELS THAT HE HAS BEEN "HIGH-BALLED" OR "LOWBALLED" CAN BECOME A CONSIDER-ABLE LIABILITY TO YOUR PRACTICE. DON'T EVEN THINK OF IT!

Perhaps the most dangerous aspect of all this is the fact that, in engineering, the same words and phrases can mean different things to different people. Thus, the consultant can, in perfectly good faith, offer to do one thing, and the client can accept, expecting to receive something different.

For example, suppose that you agree to "investigate the concept of the . . . and provide a breadboard circuit to demonstrate the principle." Now breadboard models of electronic circuits are frequently operated by lab-type power supplies. Will the client expect that what he receives will have power supplies so that it can be simply plugged into an outlet and turned on? Or will he expect to see a collection of perforated boards with wires trailing out labeled (+5V), (- 12V), (GND)?

Take the part about demonstrating the principle; that can cover a multitude of sins also. Suppose that the device in question is a sensor to monitor the humidity in a greenhouse intended for eventual incorporation into an automatic atmospheric control system. If you could show that your breadboard would provide a signal when the humidity was above the setpoint and a different signal when the humidity was below the setpoint, would the principle be proved? On the other hand, would the client be justified in making you prove that the unit would work just as effectively at any temperature between 45°F and 110°F? How about the effects of aging? Would he be justified in making you test the unit in a temperature-cycled atmosphere for 60 days or a year to prove that there was no significant degradation in its accuracy?

My point is that two reasonable people could read the sentence and come to conclusions that would change the cost of doing the job by an order of magnitude. If long life is a basic feature and freedom from temperature variation of the readings is an essential feature, then those two items should be worked into the original estimate and priced accordingly. Obviously, the cost of testing a device for six months with the possibility that it could fail and require the design and test of a second device for six months is going to be a great deal different from that of simply showing that the reading goes up and down with humidity.

In sales to the military one usually finds that each and every aspect of the desired product has been spelled out in exquisite detail with tests specified to prove that the product does indeed meet the specification. I am not suggesting that the development program for a greenhouse humidity sensor should be covered by a MIL-Spec document of 1,400 pages. However, the cardinal points of the requirement should be ascertained and agreed upon by both parties and then covered in print in the agreement.

In less hardware-oriented programs, the principle still applies. If you are to provide a report, how much of a report? A letter? 10 pages? 20 pages? If you are expected to visit certain locations, how many times and how many days per visit? If you are to supply a software program, at what point is the software program to be considered finished?

If it seems that I am stressing the obvious, I would note that this is an area in which a great deal of difficulty is encountered by consultants. The causes stem from the fact that people often employ a consultant for reasons of convenience. It is easy to make the arrangements with the consultant, and a program can be quickly set up. The arrangements can be relatively informal, with the association on a personal basis. Unfortunately, this informality is the principal source of the misunderstandings that sometimes arise.

A secondary problem can arise from your frequently working with people whose experience does not extend into your field of expertise. A maker of greenhouse controls may not know that a humidity sensor could also be temperature-sensitive or could lose sensitivity because of accumulation of mold or fungus over a six-month period. A banker may have nothing in his experience to tell him that a computer memory may be pattern-sensitive so that errors do not show up until the memory begins to fill up. When dealing with clients who have limited sophistication in your field, it is your responsibility to see that the proper things are asked for and supplied.

In this latter case, the program will frequently turn out to be longer and more expensive than the prospective client had anticipated. It is the general case that people inexperienced in product development will underestimate the effort required by a considerable margin. For that matter, people *very* experienced in development programs sometimes underestimate the effort. As noted earlier, the client will often want to have some portions of the program deleted. At this point, it is important that all parties understand what has been deleted and what remains. This ought to be spelled out in writing, since these things have a way of being forgotten during the course of the program.

As an example of the way in which the greenhouse sensor could have been handled, the sentences could have been expanded along these lines:

... The object of the investigation shall be to determine the feasibility of the use of a conductivity-type humidity sensor for eventual incorporation into an automatic humidity control system. The breadboard model will consist of hand-etched printed circuit boards mounted in edge-card connectors. The unit shall be powered using lab-type power supplies not furnished. Engineering sketches of the final circuitry shall be provided . . . . The unit will supply standard TTL level output signals to indicate:

a. Humidity above setpoint
b. Humidity below setpoint

Absence of either signal shall indicate humidity within the deadband. The deadband will be arranged to be adjustable to ±10% of the setpoint value. The setpoint will be arranged to be adjustable from 15% RH to 80%. No aging tests are included in this phase of the program.

A major portion of the investigation will be directed toward temperature compensation of the sensor. The design goal shall be to provide a variation of no more than ±3% variation in humidity reading with temperature over the range from 40°F to 110°F. A variation of ±8% RH over the temperature range shall be the minimum acceptable. . . .

In the event that any of the circuitry or techniques developed in the course this contract . . . assign sole and exclusive rights . . . . If the preparation of the patent application requires additional effort on the part of the consultant, such services shall be provided at the rate of $——/day, plus vouchered travel expenses if required, for up to 5 days' services as requested by the client.

The final engineering report shall contain . . . .

Delivery will be . . . .

Terms and conditions . . . .

This quotation is valid for 90 days from the date of submission.

While the above example is not all-inclusive, you can see that an attempt is made to nail down most of the areas that could be a source of discussion.

Summary

Always try to nail down all factors involved in the task. Never accept or write an agreement that does not clearly state what must be provided or completed for the job to be considered complete. Always specify conditions and means of payment. Failure to attend to these points can yield hurt feelings in the best case and financial ruin in the worst!

Liability

Concern over liability comes in two forms. A few firms may require that you show proof of disability and accident insurance before they will permit you or an employee on the premises. Contact your insurance agent about this. The insurance is not terribly expensive. Product liability is another matter. If you intend to produce anything that will be used by the public or by people other than knowledgeable qualified people, DISCUSS THE MATTER WITH YOUR ATTORNEY.

If, on the other hand, you are going to produce breadboard or experimental apparatus for evaluation, and it is to be handled *only* by knowledgeable engineers or trained technicians, your liability is somewhat less. However, this does not release you from the requirement to provide as much protection from high voltages, high pressures, corrosive materials and rotating parts as is consistent with ability to perform the required tests and measurements. If there is any question of exposure to potentially dangerous conditions during the course of the tests, it may be advisable to add a statement to the agreement noting the unavoidable hazard and stating that the equipment is to be operated and tested only by qualified engineers and technicians. TALK TO YOUR ATTORNEY!

The Contract

In most cases, the actual contract for the work will take the form of a Purchase Order with a serial number and a date. Usually a statement of work (SOW) will be referred to and attached to the purchase order. The latter will usually contain the agreement describing the work to be done. In preparing an invoice for the work, always refer to the serial number and date of the Purchase Order and the SOW. This information will assist the Accounts Payable people in identifying and validating the invoice and will save you a good deal of time in collecting. In cases where the project is long, the customer may agree to progress payments, which are usually tied to some event, such as delivery of the monthly progress report. Always note the actual accomplishment of the event on the invoice, indicating that the report has been sent to Mr.—— of the XYZ Corporation—Engineering Division. This gives Accounts Payable a specific individual to contact for assurance that the work has been performed.

Your being accurate, consise and businesslike gives these people a good feeling and speeds up collections considerably!

5

How Will I Finance It?

The financing of a practice or a small business is obviously a matter of prime importance to the would-be consultant or engineer. Unless he is the purest optimist, he will have to consider how he is to survive the transition from a regular payroll and paycheck to the uncertainties of self-employment. And unless he inherited money, married it or has a spouse with a substantial income, he will have to consider just how much of an investment he will have to commit to the venture—in other words, how much savings will he have to have available in order to make the grade without going broke.

Furthermore, engineers being what they are, he will no doubt wonder about the optimum point in his career to undertake such an adventure, at least from a financial point of view. This is really a pretty sophisticated question. A mathematical model is presented in the form of computer printouts and will be discussed later. It is intended to shed some light on the subject and provide some objective financial background and projections.

In the medical and dental professions, it is not uncommon to find that the new practitioner simply buys out an established practice. This arrangement has some obvious advantages, since the new practitioner can expect to keep some significant fraction of the old patients. He therefore will have some immediate income and will be speeded on the path to financial self-sustenance. Of course, he will probably have to borrow the money required for the purchase of the practice, and that will be represented in his budget as a direct expense for mortgage principal and interest payments.

Needless to say, it is difficult to find a physician or dentist who is willing to discuss the purchase price of his practice or the rate at which he could recover the investment and the capital cost thereof. However, we can make a few simplifying judgments and arrive at some figures that give a fair price for such an exchange.

To begin with, let us consider the value of a retirement fund. Let us suppose that the retiree deposits $10 in a savings bank at the age of 65 and wishes to withdraw money at the rate of $1/year until his death. Table 5-1 show how long the money would last if

Table 5-1

7%		5.5%	4%
1.	Year	1.	1.
9.63	Balance	9.49	9.36
2.		2.	2.
9.23		8.96	8.69
3.		3.	3.
8.81		8.40	8.00
4.		4.	4.
8.35		7.80	7.28
5.		5.	5.
7.87		7.18	6.53
6.		6.	6.
7.35		6.52	5.75
7.		7.	7.
6.79		5.82	4.94
8.		8.	8.
6.20		5.09	4.10
9.		9.	9.
5.56		4.31	3.22
10.		10.	10.
4.88		3.49	2.31
11.		11.	11.
4.16		2.63	1.36
12.		12.	12.
3.38		1.72	0.38
13.		13.	13.
2.54		0.76	−0.64
14.		14.	14.
1.65		−0.24	−1.70
15.		15.	
0.70		−1.31	
16.			
−0.31			

the bank paid the caption rates compounded annually. It may be seen that the $10-deposit would provide a dollar a year for 12 years at 4%, nearly 14 years at 5.5% and nearly 16 years at 7%. At 11.2%, it would provide a dollar a year in perpetuity.

If the retiree's normal income had been $2 per year, a deposit of five times his normal income would give him a 50% retirement pay for the rest of his expected life, if invested at 7% per year.

Of course, the actual figures for any given transaction are the subject of negotiations between the parties involved and would depend upon the nature of the practice and whether it was growing or declining, the neighborhood, and so forth. However, it seems fair to say that a value of five times net income plus fair-market-value for physical plant and inventory would be a reasonable point on which to base negotiations. This would at least give the retiree a reasonable retirement benefit from the transaction.

Next, let us investigate this transaction from the standpoint of the purchaser. Let us suppose that he has to borrow the money and repay a mortgage on the amount. We shall include in the bargain a 10% increment for physical plant and inventory; thus if the purchase price were $10, the buyer would have to borrow $11. The figures in Table 5-2 would apply.

Table 5-2. For $11 Borrowed,

Mortgage rate	Annual Payment for Stated Rate and Term			
	10 years	15 years	20 years	25 years
8%	$1.60	$1.26	$1.10	$1.02
10%	1.74	1.42	1.27	1.20
12%	1.89	1.58	1.45	1.39

If we presume that 90% of the patients would stay with the buyer, then the normalized income would be $1.80 per year at the outset. If the new practitioner could obtain the 8%, fifteen-year or the 10%, twenty-year mortgage, he would be able to net approximately one quarter of what the retiree had at retirement, after, presumably, 20 to 30 years to build up the practice. As we shall see, this is not too bad a deal.

Let us suppose that the practice had been only 75% saturated, that is, the retiree was doing only 75% of the practice that could be handled by the younger practitioner. If the practice grew in net income at the rate of 8% for the first five years, it would be saturated at $2.67/year. Thereafter, a 4% growth to keep up with inflation (higher rates for the same number of patients) would bring the income to $3.37/year at the end of the eleventh year. The new practitioner would be able to pay the bank payments and still have a little over $2/year left. He would be at the point the retiree had reached (at retirement) just 11 years after starting! That's not bad in anyone's book! Of course, his real purchasing power would not be as great as his predecessor's had been at that same point, owing to inflation. However, it is unrealistic to expect that a factor which has been with us since the beginning of time is likely to cease tomorrow. Also, had the inflation not been present, he could have renegotiated the loan for a much lower rate.

The ability to obtain a loan to purchase a practice such as this is more common in the medical and dental professions than it is in the consulting engineering field. Bankers have some reason to be more confident about the earning power of the physician or dentist (try orthodontist) than that of the engineer!

A second reason for the relative uncommonness of this type of transaction is the fact that self-employment is relatively uncommon in engineering compared to either dentistry or medicine. There is more question about whether the junior practitioner would be able to retain a major fraction of the clientele.

In principle, I see no reason why a consulting engineering practice could not be successfully purchased—most particularly if the purchaser were to enter the firm on a junior partner basis and gradually take over the practice. This arrangement would afford him the opportunity to meet and serve the clientele, and the transfer could be a smooth and orderly procedure. The few sales of private engineering practices probably proceed in this manner. Very few factual data are available on the subject.

ENGINEERING INCOMES

In order to understand the following treatment we must first view a few models of an engineer's income and the way it varies with time, in order to arrive at some reasonable conclusions about when a person might be ready to try a private practice and what economic factors tend to affect the choice.

The model employed in this study was developed by the writer as a strictly mathematical artifice, an empirical curve-fitting exercise to provide figures that fit the available past data. This is the "Farmers' Almanac" approach to prediction of the future. If records indicate that it has snowed in the Northeast on fourteen of the last twenty March 16ths, then it seems advisable to predict snow for the next March 16. The data presented have been garnered from personal records, salary surveys presented in trade magazines and data available from other popular sources. No causal economic theory is presented.

Perhaps the first and foremost factor is the inflation factor used in the predictions. It will be noted from the equations used in the calculations that the inflation rate enters into both the model for the salary and the income tax, and thereby into the savings equation as well.

The actual value for this inflation factor is a subject of considerable debate among economists and politicians, and your humble servant is probably not qualified in either of their fields. The selected inflation rate of 4% seems to fit a series of items fairly well. Between the years 1950 and 1977, it implies a multiplication factor of 2.883 in prices. The apartment that rented for $75 in 1950 goes for about $216 in 1977. Similarly, the $3,600 starting salary for engineers of 1950 is translated to $10,380 in 1977. The $1,600 Ford Tudor becomes $4,613, and so on. I am sure that some would quibble about the exact rate; however, it would seem that 4% represents a reasonable engineering approximation.

On the subject of the engineer's salary, there is a certain amount of economic theory presented. At the time of initial hiring, the neophyte's salary is relatively low; the initial increases come rather rapidly and are prone to be rather large. By the time that the engineer has been out of school for 10 years or so, his rate of annual increase has fallen rather considerably, although the dollar amounts are actually higher. By the time that he has been out of school 30 years and he is about 52 years of age, the percentage of increase has fallen to approach the inflation rate. His last few salary increases before retirement will scarcely be above the inflation rate.

If you stop to think about it, this makes a certain amount of sense in terms of the value of the engineer to the company. If we made the presumption that his compensation is directly proportional to his value to the company, we would arrive at a similar relationship. At the time of his first hiring, he will be of relatively small value to the company. If he is a good performer, his value will be increasing dramatically with each year. By the 10 year mark, he will have matured and should be a journeyman in his art. His value will be increasing more slowly because it is already fairly high. By the time he is 60, another year of experience will scarcely increase his value at all. This effect is reflected in Table 5-3, for three classes of men.

Table 5-3

Initial rate of increase	OUTSTANDING 13% above inflation	HIGH AVERAGE 9%	INFLATION ONLY 0%
After — yr			
5	11.9% absolute	9.5%	4%
10	8.8%	7.3%	4%
15	6.9%	6.0%	4%
20	5.8%	5.2%	4%
30	4.6%	4.4%	4%
43	4.2%	4.1%	4%

The data taken for these curves came from available salary surveys, which included samples from electrical, mechanical and aeronautical engineers principally, with some smattering of others. The data were normalized for inflation rate in the process of reduction and curve fitting. A word of explanation is also in order on the categorization.

The performance categorized as Outstanding represents the top percentile of the curves. A typical example is the person who was a Project Engineer at the 6-year mark, a Section Supervisor at the 10- or 11-year mark and a Manager with perhaps five Section Supervisors reporting to him at the 15- to 17-year mark. Alternatively, he might have been a Consultant or Senior Staff Engineer at the 15-year mark. At retirement he might be a Division President or Senior Scientist.

The High Average performer probably became a Project Engineer at the 10- to 12-year mark and a Section Supervisor at the 20-year mark. At retirement he may be a manager with perhaps five Section Supervisors reporting to him, or he might be a Consultant or Senior Staff Engineer.

The column headed Inflation Only probably doesn't represent anyone. The sorry soul who couldn't do any better than that would probably have been fired for incompetence long before he reached the 10-year mark. The column is included only for a benchmark of the inflation in subsequent data.

The man in the Outstanding column may have had to change jobs at the 5- to 7-year mark, again at the 12- to 14-year mark and perhaps again at the 20-year mark in order to achieve this performance. The High Average man could perhaps achieve the stated performance with a single job change.

Data included in the curve-fitting process represent people with graduation dates ranging from 1914 through 1975; however, the overwhelming majority of the data comes from people who graduated after 1948 because most engineers have graduated since then. (Note that I said *most engineers*, not just most active or living engineers.) The magazine and organization salary surveys are subject to interpretation, since they represent only the engineers who were willing to respond. They are presented not as average or even necessarily representative figures. In general, the people seriously considering entrepreneurship or consulting practice will be well above average in ability and earning power. It is these people with whom we are concerned. The Outstanding and High Average curves probably cover this range reasonably well. (See Table 5-4.)

A second objection could perhaps be raised in view of the fact that the principal period covered is the one from 1950 to 1977. The 1949–1951 period marked a time of very poor engineering employment opportunity, owing to the glut of World War II-G.I. Bill graduates. The period from 1970 to 1975 also represents a low period. Conversely, the periods 1952–1958 and 1960–1967

Table 5-4

OUTSTANDING

```
        0 • 1 3 0 0
 • • • • • • • • • • •
```

Start yr   1,950•
Start pay   3,600•00
Svgs   0•00

Yr 1,951•	1,957•	1,963•
Pay 4,170•35	8,548•08	14,522•31
Fed Tax 371•25	1,054•00	2,230•13
Σ F.I.T. 371•25	4,772•42	14,946•51
Σ Income 3,228•74	34,013•35	85,267•83
Svgs Dep 96•24 Dep	195•43	320•45
Σ Svgs 96•24	1,143•55	3,326•24
1,952•	1,958•	1,964•
4,787•38	9,431•78	15,683•82
454•92	1,212•97	2,482•76
826•18	5,985•40	17,429•28
6,944•17	41,348•45	101,307•37
110•51	214•77	342•93
211•57	1,415•50	3,835•48
1,953•	1,959•	1,965•
5,450•37	10,359•19	16,897•90
550•03	1,386•00	2,753•86
1,376•21	7,371•41	20,183•14
11,181•51	49,394•23	114,237•33
125•76	234•77	365•58
347•91	1,721•06	4,392•84
1,954•	1,960•	1,966•
6,158•55	11,330•86	18,167•34
657•07	1,573•61	3,044•55
2,033•29	8,945•02	23,227•70
15,974•81	58,179•81	128,090•68
141•93	255•40	388•13
507•24	2,062•51	5,000•62
1,955•	1,961•	1,967•
6,911•24	12,347•68	19,495•31
776•47	1,776•38	3,356•13
2,809•77	10,721•41	26,583•83
21,356•89	67,734•29	142,901•90
158•96	276•59	410•07
691•57	2,442•24	5,660•73
1,956•	1,962•	1,968•
7,707•86	13,410•95	20,885•28
908•64	1,994•96	3,690•01
3,718•42	12,716•37	30,273•84
27,359•48	78,087•01	158,707•20
176•91	298•30	430•23
902•97	2,862•65	6,374•00

OUTSTANDING

1,969 •	1,975 •	4,981 •
22,340 • 91	32,215 • 73	45,681 • 22
4,047 • 78	6,686 • 64	14,696 • 40
34,321 • 63	67,368 • 47	120,665 • 23
175,544 • 69	299,846 • 93	474,359 • 48
444 • 91	669 • 40	945 • 97
7,137 • 62	14,498 • 88	19,639 • 71
1,970 •	1,976 •	4,982 •
23,798 • 71	34,167 • 48	48,399 • 16
4,431 • 15	7,239 • 30	11,557 • 51
38,752 • 78	74,607 • 78	132,222 • 74
193,454 • 45	324,823 • 36	505,483 • 19
-1,790 • 97	711 • 42	999 • 14
5,703 • 52	14,735 • 25	24,620 • 84
1,971 •	1,977 •	4,983 •
25,324 • 22	36,224 • 99	51,279 • 89
4,823 • 44	7,832 • 96	12,487 • 47
43,576 • 23	82,440 • 75	144,710 • 22
212,429 • 72	351,157 • 88	541,394 • 88
501 • 40	754 • 76	1,054 • 79
6,490 • 10	13,076 • 78	23,756 • 68
1,972 •	1,978 •	4,984 •
26,922 • 81	38,396 • 27	54,335 • 30
5,242 • 70	8,471 • 30	13,492 • 56
48,618 • 93	90,912 • 05	158,202 • 78
232,511 • 24	378,911 • 56	579,182 • 20
546 • 63	799 • 70	1,113 • 10
7,361 • 25	14,530 • 32	26,057 • 62
1,973 •	1,979 •	1,985 •
28,600 • 14	40,689 • 87	57,578 • 05
5,691 • 25	9,158 • 33	14,579 • 63
54,510 • 19	100,070 • 39	172,782 • 42
253,742 • 79	408,149 • 51	618,937 • 87
587 • 83	846 • 42	1,174 • 23
8,317 • 15	16,103 • 26	28,534 • 74
1,974 •	1,980 •	1,986 •
30,362 • 27	43,114 • 93	61,021 • 69
6,171 • 64	9,898 • 43	15,756 • 14
60,681 • 83	109,968 • 82	188,538 • 57
276,171 • 29	438,940 • 95	660,759 • 78
628 • 39	895 • 12	1,238 • 36
9,361 • 40	17,803 • 55	31,199 • 85

HIGH AVERAGE

```
             0.09
.............
           1,950.
         3,600.00
             0.00
```

Yr	1,951.	1,963.	1,957.
Pay	4,040.05	11,363.33	7,211.65
Fed Tax	372.21	1,615.69	854.65
Σ F.I.T.	372.21	11,809.25	4,172.75
Σ Income	3,227.78	76,637.25	34,154.01
Svgs Dep	96.21	257.29	169.32
Σ Svgs	96.21	2,889.48	1,052.35
	1,952.	1,964.	1,958.
	4,505.59	12,169.27	7,830.40
	435.19	1,776.21	960.15
	807.41	13,585.46	5,132.90
	6,832.64	86,224.37	37,405.51
	107.22	273.07	183.04
	208.25	3,307.03	1,288.01
	1,953.	1,965.	1,959.
	4,996.34	13,013.26	8,476.62
	505.50	1,948.16	1,073.58
	1,312.92	15,533.63	6,206.49
	10,832.73	96,445.47	44,162.33
	118.72	288.99	197.16
	337.38	3,761.38	1,549.57
	1,954.	1,966.	1,960.
	5,512.17	13,897.88	9,151.50
	582.39	2,132.41	1,195.34
	1,895.31	17,666.05	7,401.83
	15,246.68	107,326.31	51,443.61
	130.69	304.85	211.67
	484.95	4,254.30	1,838.72
	1,955.	1,967.	1,961.
	6,053.12	14,825.91	9,856.44
	666.05	2,329.93	1,325.90
	2,561.37	19,995.98	8,727.74
	20,092.80	118,894.26	59,269.21
	143.13	320.28	226.54
	652.34	4,787.30	2,157.21
	1,956.	1,968.	1,962.
	6,619.46	15,800.35	10,593.09
	756.72	2,541.76	1,465.82
	3,318.09	22,537.74	10,193.56
	25,389.20	131,178.42	67,659.84
	156.01	334.38	241.76
	840.97	5,361.05	2,506.84

HIGH AVERAGE

1,969•	1,975•	1,981•
16,824•25	23,835•40	33,602•04
2,769•06	4,453•93	7,059•65
25,306•81	47,503•67	82,767•11
144,209•71	239,426•19	369,942•54
344•34	510•45	719•80
5,973•44	6,706•04	15,905•81
1,970•	1,976•	1,982•
17,848•86	25,238•64	35,593•25
3,013•08	4,810•06	7,624•74
28,319•89	52,313•73	90,391•86
158,020•88	258,451•52	395,919•64
-1,381•11	541•91	760•61
4,891•00	9,683•25	17,461•72
1,971•	1,977•	1,983•
18,925•18	26,723•10	37,709•28
3,262•17	5,193•82	8,236•73
31,582•06	57,507•56	98,628•59
172,607•59	278,496•34	423,276•35
385•44	574•50	803•51
5,520•99	10,741•92	19,138•32
1,972•	1,978•	1,984•
24,057•40	28,294•90	39,959•23
3,529•09	5,607•75	8,899•95
35,111•15	63,115•32	107,528•55
186,003•69	299,611•69	452,085•68
419•09	608•42	848•64
6,216•14	11,887•44	20,943•88
1,973•	1,979•	1,985•
21,249•97	29,960•59	42,352•81
3,815•47	6,054•63	9,619•14
38,926•63	69,169•96	117,147•69
204,245•61	321,851•96	482,425•77
449•69	643•84	896•13
6,976•63	13,125•66	22,887•21
1,974•	1,980•	1,986•
22,507•63	31,727•14	44,900•41
4,123•10	6,537•49	10,399•49
43,049•74	75,707•45	127,547•19
221,372•48	345,275•05	514,379•09
479•85	680•92	946•15
7,805•32	14,462•86	24,977•72

INFLATION ONLY

```
                    0 • 0 0
          • • • • • • • • • •

                  1,9 5 0 •
            3,6 0 0 • 0 0
                  0 • 0 0
```

Yr	1,9 5 1 •	1,9 5 7 •	1,9 6 3 •
Pay	3,7 4 6 • 8 8	4,8 4 1 • 1 9	6,4 2 9 • 1 5
Fed Tax	3 7 1 • 2 5	5 1 9 • 8 1	7 5 6 • 4 9
Σ Tax	3 7 1 • 2 5	3,0 8 6 • 1 1	6,9 7 8 • 8 0
Σ Income	3,2 2 8 • 7 4	2 5,5 3 7 • 0 2	5 4,3 7 0 • 2 1
Svgs dep	9 6 • 2 4	1 2 0 • 7 1	1 5 3 • 7 5
Σ Svgs	9 6 • 2 4	8 7 2 • 4 5	2,1 1 0 • 8 8

1,9 5 2 •	1,9 5 8 •	1,9 6 4 •
3,9 0 2 • 7 8	5,0 6 6 • 0 9	6,7 5 7 • 6 1
3 9 1 • 6 1	5 5 1 • 8 9	8 0 3 • 2 2
7 6 2 • 8 6	3,6 3 8 • 0 1	7,7 8 7 • 0 2
6,5 8 4 • 0 2	2 9,8 2 6 • 3 2	5 9,9 9 1 • 1 4
9 9 • 8 0	1 2 5 • 5 9	1 6 0 • 1 0
2 0 0 • 8 5	1,0 4 1 • 6 7	2,3 7 6 • 5 3

1,9 5 3 •	1,9 5 9 •	1,9 6 5 •
4,0 6 8 • 3 3	5,3 0 5 • 4 7	7,1 0 7 • 7 7
4 1 3 • 5 3	5 8 6 • 5 9	8 6 4 • 3 3
1,1 7 6 • 3 9	4,2 2 4 • 6 0	8,6 5 1 • 3 6
1 0,0 7 3 • 2 7	3 4,3 0 5 • 8 1	6 5,8 8 4 • 4 1
1 0 3 • 5 6	1 3 0 • 7 1	1 6 6 • 6 2
3 1 4 • 4 6	1,2 2 4 • 4 6	2,6 6 1 • 9 9

1,9 5 4 •	1,9 6 0 •	1,9 6 6 •
4,2 4 4 • 1 9	5,5 6 0 • 3 4	7,4 8 1 • 1 3
4 3 7 • 1 5	6 2 4 • 1 3	9 2 5 • 2 3
1,6 1 3 • 5 5	4,8 4 3 • 7 4	9,5 7 6 • 5 9
1 3,7 0 4 • 4 5	3 8,9 8 7 • 1 5	7 2,0 6 6 • 9 6
1 0 7 • 5 2	1 3 6 • 0 8	1 7 3 • 2 2
4 3 7 • 7 0	1,4 2 1 • 7 8	2,9 6 8 • 3 1

1,9 5 5 •	1,9 6 1 •	1,9 6 7 •
4,4 3 1 • 1 0	5,8 3 1 • 7 9	7,8 7 9 • 2 7
4 6 2 • 6 2	6 6 4 • 7 7	9 9 1 • 3 4
2,0 7 6 • 1 7	5,5 1 3 • 5 1	1 0,5 6 7 • 9 4
1 7,4 8 6 • 0 2	4 3,8 8 2 • 7 2	7 8,5 5 6 • 7 5
1 1 1 • 6 9	1 4 1 • 7 2	1 7 9 • 6 8
5 7 1 • 2 8	1,6 3 4 • 5 9	3,2 9 6 • 4 1

1,9 5 6 •	1,9 6 2 •	1,9 6 8 •
4,6 2 9 • 8 2	6,1 2 0 • 9 8	8,3 0 3 • 8 4
4 9 0 • 1 2	7 0 8 • 7 9	1,0 6 3 • 1 4
2,5 6 6 • 3 0	6,2 2 2 • 3 0	1 1,6 3 1 • 0 9
2 1,4 2 7 • 0 1	4 9,0 0 5 • 7 3	8 5,3 7 2 • 8 7
1 1 6 • 0 9	1 4 7 • 6 1	1 8 5 • 5 4
7 1 5 • 9 4	1,8 6 3 • 9 4	3,6 4 6 • 7 7

INFLATION ONLY

1969.	1975.	1981.
8,756•55	11,973•54	16,734•42
1,141•14	1,738•28	2,717•62
12,772•23	21,580•56	35,195•37
92,535•58	143,707•54	212,920•00
189•26	271•68	381•85
4,018•38	5,745•05	9,946•85

1970.	1976.	1982.
9,210•80	12,642•08	17,729•24
1,225•86	1,868•88	2,935•72
13,998•10	23,449•45	38,131•10
100,066•27	153,812•20	226,718•69
-753•06	287•81	404•02
3,466•23	6,320•12	10,843•22

1971.	1977.	1983.
9,694•68	13,355•99	18,792•95
1,312•27	2,010•97	3,173•61
15,310•38	25,460•42	41,304•72
107,964•79	164,443•31	241,274•32
208•71	304•69	427•52
3,848•25	6,940•83	11,818•16

1972.	1978.	1984.
10,210•45	14,118•54	19,930•52
1,405•96	2,165•64	3,433•19
16,716•36	27,626•06	44,737•91
116,253•49	175,633•67	256,634•09
225•62	322•44	452•45
4,266•29	7,610•31	12,861•52

1973.	1979.	1985.
10,760•49	14,933•24	21,147•24
1,507•70	2,334•08	3,716•50
18,224•07	29,960•14	48,454•42
124,956•24	187,418•13	272,848•10
240•95	341•15	478•90
4,720•56	8,331•98	13,983•50

1974.	1980.	1986.
11,347•29	15,803•86	22,448•82
1,618•20	2,517•60	4,025•83
19,842•28	32,477•75	52,480•25
134,098•53	199,833•77	289,969•52
256•14	360•93	506•97
5,212•73	9,109•51	15,189•65

GERIATRIC SUMMARY

0.13	0.09	0.00
1,987•	1,987•	1,987•
64,680•65	47,613•10	23,841•35
17,030•26	14,246•67	4,363•65
205,568•94	138,793•87	56,843•91
704,751•21	548,032•83	308,054•69
1,305•67	998•84	536•77
34,065•52	27,225•46	16,485•90
1,988•	1,988•	1,988•
66,570•34	50,502•72	25,331•37
18,410•91	12,166•86	4,732•69
223,979•75	154,960•73	61,576•60
751,020•95	583,479•07	327,163•35
1,376•33	1,054•37	568•40
37,145•12	29,641•11	17,878•60
1,989•	1,989•	1,989•
72,707•18	53,581•91	26,925•86
19,907•79	13,166•83	5,135•94
243,887•55	164,127•57	66,712•54
799,683•50	620,814•95	347,358•78
1,450•52	1,112•90	601•98
40,452•91	32,236•07	19,374•51
1,990•	1,990•	1,990•
77,108•70	56,864•19	28,632•32
21,531•55	14,253•99	5,576•66
205,419•10	178,381•57	72,289•21
850,859•12	660,142•88	368,707•98
1,528•45	1,174•59	637•63
44,004•01	35,022•47	20,980•87
1,991•	1,991•	1,991•
84,793•61	60,363•97	30,458•77
23,293•77	15,436•42	6,058•46
288,712•87	193,817•99	76,347•67
904,674•06	701,570•64	391,281•84
1,610•29	1,239•63	675•47
47,814•50	38,013•23	22,705•39
1,992•	1,992•	1,992•
86,781•85	64,096•66	32,413•81
25,207•09	16,722•96	6,585•25
313,919•97	210,540•96	84,932•93
961,260•59	745,211•65	415,155•35
1,696•25	1,308•19	715•63
51,901•48	41,222•09	24,556•30

GERIATRIC SUMMARY

0.13	0.09	0.00
1,993•	1,993•	1,993•
92,094•70	66,078•71	34,506•64
27,285•33	18,123•27	7,161•37
341,205•31	228,664•23	92,094•30
402,757•11	791,185•05	446,407•80
1,786•51	1,380•45	758•26
56,283•07	44,663•65	26,542•37
1,994•	1,994•	1,994•
97,754•83	72,327•69	36,747•09
29,543•56	19,647•90	7,791•53
371,743•87	248,312•14	99,885•83
1,083,308•24	835,615•86	467,122•91
1,881•29	1,456•61	803•48
60,978•52	48,353•44	28,672•98
1,995•	1,995•	1,995•
103,786•44	76,862•35	39,145•72
34,998•22	21,308•39	8,480•91
402,747•10	265,620•53	108,366•75
1,149,064•85	890,635•15	495,389•09
1,980•78	1,536•85	851•46
66,008•24	52,307•97	34,958•09

represent eras of very high demand, with a slight trough between 1958 and 1960. This is true, but the data are the only data available in any quantity. Whether it will be possible to extrapolate this material for future performance remains to be seen.

Calculation of federal income tax is based upon a curve fit to current and earlier tax tables. The calculation is based upon a married couple with one engineer only working, and with two children. It is noteworthy that the base number of $1,734, which represents the lower no-tax limit, fits the 4% inflation rate reasonably well, since it extrapolates to $5,000 in 1977, at which point the family would probably be due an earned-income credit.

The savings included in the computation represent 10% of disposable income. This figure actually includes a certain amount of retirement benefit, which usually becomes liquid at job changes accomplished before age 50. It may also represent some equity in a home, etc. If those things are included, the totals may not seem so high. There is also a damping factor included; it is the funny-looking function within the brackets. This is actually a gauss error function designed to represent the rather large jolt that occurrs at about 20 years when the two children go to college.

The summation of disposable income and income tax is presented more out of interest than from direct applicability to the problem.

The actual calculations used are shown in equation form on page 76. These were set up in a Monroe 326 Scientist and printed out in the form of an annual statement. The table of these statements is shown as Table 5-4.

For the tables, it was necessary that a common starting year be selected; it was set at 1950. For the savings, an interest rate of 5% was selected, which would have been a bit high for a savings account in 1950 but would have been low if any significant portion of the savings had been in common stock in 1977. It was assumed that the taxable income (labeled Pay) would include 3% of accumulated savings. All three of the men were started at the same $3,600/year figure in 1950. The main dip in savings comes in 1970 and is due to college costs.

In 1977, we find that the Outstanding man is earning $36k while the High Average man is near $27k, and the poor soul on Inflation Only is at $13k. The income tax for the year runs $8k, $5k and $2k. The savings account contains $13k, $11k and $7k. The relatively smaller spread in the savings compared to the earnings is due largely to the FIT Federal Income Tax), into which the three have paid respectively $82k, $58k and $25k. In reality, this is only a part of the tax differential, since state and city income taxes can apply a multiplier to this factor.

In addition, in the normal course of affairs Outstanding is likely to live in neighborhoods where his school and real estate taxes will be higher than those of Inflation Only.

While it is questionable to carry the results of this sort of extrapolation too far, it is interesting to have a look at the predictions extended to retirement time. The section of Table 5-4 labeled Geriatric Summary shows these results. Outstanding will have earned a lifetime income of 1 megabuck if he retires in 1993 and will have paid $340k in income taxes besides! For High Average

INCOME AND TAX
MODELS

Salary for the n'th year:

$$\text{Salary} = S_0 r_c{}^{(t)}$$

where:

S_0 = starting gross pay
t = time in years after start
r_c = rate of increase, determined by:

$$r_c = 1 + r_0 \epsilon^{-4.5t/45} + r_i$$

where:

r_0 = INITIAL INCREASE RATE
r_i = rate of general inflation (taken as 0.04)

$$\text{Pay} = \text{Salary} + \text{Interest income}$$
$$= S_0 r_c{}^{(t)} + 0.03 \, \Sigma \text{ savings}$$

Income tax is computed by:

$$\ln \text{F.I.T.} = \frac{4}{3} \ln (\text{Pay} - \text{Ded}) - 4.933$$

Deduction and exemption are computed on the basis of a married couple with two children:

$$\text{Ded} = (1.04)^{t_1} \times \$1,734$$

where

t_1 = years after 1950

$$\text{Disposable Income} = \text{Pay} - \text{F.I.T.}$$
$$\text{Savings Deposit} =$$
$$\text{Dep} = 0.10(\text{Pay} - \text{F.I.T.}) \left[1 - 2\epsilon^{-\left(\sqrt[32]{(t-20)^2}\right)} \right]$$

$$\text{Savings Increment} = \text{Dep} + 0.05 \, \Sigma \text{ savings}_{(n-1)}$$

the numbers are \$790k and \$228k, respectively. For Inflation Only, the numbers are \$440k and \$92k. Outstanding will have outperformed Inflation Only by a factor of 2.3 in disposable earnings and will have paid 3.7 times as much income tax. His savings will be greater by only 2.12.

FUNDING THE BUSINESS

Now that you have had an opportunity to look over the tables and to speculate about the validity of the model and perhaps try to figure out where you and some of your cohorts fit in the scheme of things, we shall get around to the subject we started on; namely the financing of your own proprietorship or consulting practice. Chances are that you did not graduate in 1950; therefore, the model has been normalized to starting pay and years after graduation (Table 5-5). The actual amounts of cash are firmly fixed in time by inflation; therefore, they have been deleted. The significant entry here is the amount of savings expressed in terms of current income. Table 5-5 shows these results and concludes with a Geriatric Summary. The presentation of accumulated savings as a fraction of current income is intended to reflect Parkinson's Income Law, which states that "Expenses rise to meet income." This may not be one of the immutable laws of the universe, but it seems to be true that one adjusts his standard of living to match his income. Furthermore, a person usually has some difficulty backing down. The house mortgage payments, the orthodontist's bill and the car insurance are not about to decrease dramatically when he stops receiving a weekly paycheck.

An interesting point is that Outstanding took 35 years after graduation to reach a point where he had six months' income represented in savings (remember, including part of his pension fund). High Average managed this feat in 33 years, and Inflation Only in 26. Now admittedly, Outstanding has more bucks in the bank; but chances are, he also has a larger list of relatively irreducible expenses, such as a house mortgage, insurance, and so on.

The curves in Figures 5-1 and 5-2 are intended as an aid to visualizing this spread. Figure 5-1 shows just about what we expect. In terms of normalized pay, that is, pay related to starting pay, Outstanding is above the others and is rising at a faster rate in terms of total dollars. His standard of living is not really increasing as fast as the curves would seem to indicate, as is shown by the curve of normalized pay for Outstanding divided by normalized pay for Inflation Only. As a matter of fact, it may be seen that there is even a slight droop after 35 years, due to the fact that Outstanding

Table 5-5

OUTSTANDING

```
   0 • 13
• • • • • • • • • • •
```

0 • Years	11 •	22 •
1•00 Pay	3•4299	7•4785
0•00 Σ Svgs / Pay	0•1977	0•2734
1 •	12 •	23 •
1•1584	3•7252	7•9444
0•0230	0•2134	0•2908
2 •	13 •	24 •
1•3298	4•0339	8•4339
0•0441	0•2290	0•3083
3 •	14 •	25 •
1•5139	4•3566	8•9488
0•0638	0•2445	0•3258
4 •	15 •	26 •
1•7107	4•6938	9•4909
0•0823	0•2599	0•3434
5 •	16 •	27 •
1•9197	5•0464	10•0624
0•1000	0•2752	0•3609
6 •	17 •	28 •
2•1410	5•4153	10•6656
0•1171	0•2903	0•3784
7 •	18 •	29 •
2•3744	5•8014	11•3027
0•1337	0•3051	0•3957
8 •	19 •	30 •
2•6199	6•2058	11•9763
0•1500	0•3194	0•4129
9 •	20 •	31 •
2•8775	6•6107	12•6892
0•1661	0•2396	0•4299
10 •	21 •	32 •
3•1474	7•0345	13•4442
0•1820	0•2562	0•4467

HIGH AVERAGE

0•0900
••••••••••

0 • Years	11 •	22 •
1•00 Pay	2•7379	5•5715
0•00 Σ Svgs	0•2188	0•3099
Pay		

1 •	12 •	23 •
1•1222	2•9425	5•9027
0•0238	0•2366	0•3283

2 •	13 •	24 •
1•2515	3•1564	6•2521
0•0462	0•2542	0•3467

3 •	14 •	25 •
1•3878	3•3803	6•6209
0•0675	0•2717	0•3652

4 •	15 •	26 •
1•5311	3•6148	7•0107
0•0879	0•2890	0•3836

5 •	16 •	27 •
1•6814	3•8605	7•4231
0•1077	0•3061	0•4019

6 •	17 •	28 •
1•8387	4•1183	7•8597
0•1270	0•3229	0•4201

7 •	18 •	29 •
2•0032	4•3889	8•3224
0•1459	0•3393	0•4381

8 •	19 •	30 •
2•1751	4•6734	8•8131
0•1644	0•3550	0•4558

9 •	20 •	31 •
2•3546	4•9580	9•3339
0•1828	0•2740	0•4733

10 •	21 •	32 •
2•5420	5•2570	9•8870
0•2009	0•2917	0•4905

INFLATION ONLY

0•0000
••••••••••

	Years	11•	22•
0•	Pay	1•6199	2•8362
1•00	Σ Svgs	0•2802	0•4178
0•00	Pay		

1•
1•0408
0•0256

12•
1•7002
0•3045

23•
2•9890
0•4386

2•
1•0841
0•0514

13•
1•7858
0•3283

24•
3•1520
0•4593

3•
1•1300
0•0772

14•
1•8771
0•3516

25•
3•3259
0•4798

4•
1•1789
0•1031

15•
1•9743
0•3745

26•
3•5116
0•4999

5•
1•2308
0•1289

16•
2•0780
0•3967

27•
3•7099
0•5196

6•
1•2860
0•1546

17•
2•1886
0•4183

28•
3•9218
0•5390

7•
1•3447
0•1802

18•
2•3066
0•4391

29•
4•1481
0•5579

8•
1•4072
0•2056

19•
2•4323
0•4588

30•
4•3899
0•5764

9•
1•4737
0•2307

20•
2•5585
0•3763

31•
4•6484
0•5943

10•
1•5445
0•2557

21•
2•6929
0•3969

32•
4•9247
0•6118

GERIATRIC SUMMARY

0.13	0.09	0.00
33•	33•	33•
14•2444	10•4748	5•2202
0•4632	0•5075	0•6288
34•	34•	34•
15•0931	11•0998	5•5362
0•4795	0•5241	0•6453
35•	35•	35•
15•9939	11•7647	5•8742
0•4955	0•5403	0•6612
36•	36•	36•
16•9504	12•4723	6•2357
0•5112	0•5562	0•6766
37•	37•	37•
17•9668	13•2259	6•6225
0•5266	0•5718	0•6914
38•	38•	38•
19•0473	14•0285	7•0364
0•5417	0•5869	0•7057
39•	39•	39•
20•1964	14•8839	7•4794
0•5563	0•6016	0•7195
40•	40•	40•
21•4190	15•7956	7•9534
0•5706	0•6158	0•7327
41•	41•	41•
22•7204	16•7678	8•4607
0•5845	0•6297	0•7454
42•	42•	42•
24•1060	17•8047	9•0038
0•5980	0•6431	0•7575
43•	43•	43•
25•5818	18•9108	9•5851
0•6111	0•6560	0•7691
44•	44•	44•
27•1541	20•0911	10•2075
0•6237	0•6685	0•7802

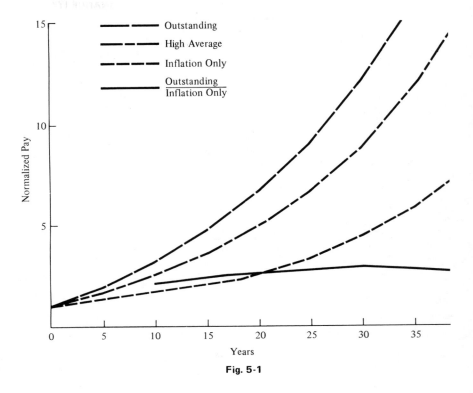

Fig. 5-1

is paying considerably more in Federal Income Tax. This rise in tax is, in fact, sufficient to overpower the rise in salary, which is actually slightly higher than inflation. Thus, for our model Outstanding would experience a small decrease in real purchasing power after 35 years.

Figure 5-2 shows another aspect of the model. In this figure, it may be seen that High Average is able to build up his normalized savings faster than Outstanding. This phenomenon is also again a function of the Federal Income Tax. It should be remembered that Outstanding will actually have more dollars in the bank, but in terms of months of income in the bank, High Average has the edge.

The offset kick in the curve shows the effect of college expense. It is presumed that each will spend money on college in proportion to pay or disposable income; therefore Outstanding is actually spending considerably more. This coupled with his higher tax rate is responsible for the fact that it will take him longer to recover from the transient of college costs. Probably in neither case would their children be eligible for any significant scholarship aid. On the other hand, Inflation Only's children could well receive nearly 100% scholarship aid. The model was not adjusted for such assistance; therefore, the curve for Inflation Only is not shown, since in his case the model is a bit misleading with respect to college costs.

In the cases of Outstanding and High Average, you will note that there is a span of about six years when savings are depressed because of college expense. Salary, taxes and disposable income are all rising during this period; however, the normalized savings re-

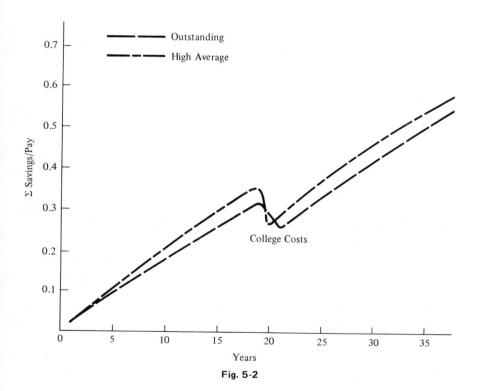

Fig. 5-2

cover slowly. This finding agrees with the common sense dictum that it would not be too smart to attempt to go into a private practice while the kids are in college. This does not mean that it cannot be done with success, but it does mean that it would be a good deal harder. I know this from personal experience.

USING THE TABLE

Let us consider an example from the table. Suppose that High Average decided to go into business for himself in 1977 after 17 years experience. His starting salary was $3,600 $(1.04)^{10}$ (for graduation in 1960) or $5,329 per year. His current pay is a little more than 4.1 times as great, say $21,946, and his savings consist of about 32% or $7,086. If he continued his employment, he would expect to pay a federal income tax, which we can calculate from the formulas given on page 76.

$$Ded = (1.04)^{t_1} \times \$1734$$
$$t_1 = 28 \text{ years since } 1950$$
$$Ded = \$5199.75$$
$$\ln \text{F.I.T} = \frac{4}{3} \ln (\text{Pay} - \text{Ded}) - 4.933$$
$$\ln \text{F.I.T} = \frac{4}{3} \ln (\$16,746.28) - 4.933$$
$$\ln \text{F.I.T} = 8.03$$
$$\text{F.I.T} = \$3086.86$$

If we were to deduct the income tax from pay we would see that our prospective On-Your-Owner has been living on a disposable income of $18,859 per year. With a state income tax that tends to run about a third of the federal tax, he has actually been living on a disposable income of about $17,830 per year, or about $1485 per month.

The business of neglecting taxes for the first year of practice is not too bad an idea, since the chances of making enough income to be required to pay any significant tax are relatively small, as we shall see shortly. One thing is eminently apparent, however, and that is that expenditures will have to be reduced drastically if the $7,086 in savings is to last long enough to keep the family in business.

You will note that the choice of the 17-year mark should put him in business and have him encounter the major losses associated with start-up something more than a year ahead of the time when the children will be going to college. There is a distinct advantage to this, since the general provision for college aid is based upon the previous one or two years' income tax reports. The savings are at a high, which they will not recover for six years. If the first two years of self-employment predeed the college enrollment, scholarship aid will be available, since for the first year or so you will generally be near or even below poverty earning level. It can just as readily work the other way. If the business is started with the kids actually in college, the previous years' salaries will positively preclude any scholarship aid at the same time that income falls to the poverty level.

There is also a powerful chastening influence on the family to have been involved in a low-budget operation before the college starts. The ski trips get to be a lot less expensive. At the start, self-employment can be a bit like begging in India—an honored but humbling profession!

6

How Long Will It Take?

The length of time it will take to get the practice established is a function of a number of things. Superficially, it can be observed that five years represents about the median time to get things going. A majority of my self-employed friends agree that *if* you are going to make it, it takes about five years to get to the point where your income is at about the level where it was before you started. Generally you will be beginning the seventh year before your spouse will start to use the charge accounts again and be willing to buy something like a chair or davenport without some specific commitment regarding the source of the money. In other words, it is that long before you are willing to spend on the expectation of income rather than firm evidence, such as a check received.

During this period, as we shall see shortly, there is usually a point where savings can dip to a distressingly low level. If the start or the middle of the seventh year sees your savings back at the pre-business level, you can feel that you have done very well.

Let us go back to Mr. High Average now, and examine some of these statements in the light of the example. We remember that he had $7,086 in the bank and his family was used to living on $1,485 per month. On the face of it, it is apparent that his standard of living is going to have to change, since these savings amount to less than 5 months worth of income. Let us examine first of all just how far he can reasonably be expected to pare expenditures during the start-up period.

To begin with, let us presume that HA and wife purchased a small home five years after graduation. They put $3,000 down and got a 6%, thirty-year mortgage on the house, which cost $20,630. Owing to inflation, the value of the house has appreciated to $33,029 in 12 years. With the 6% mortgage he was paying $6/$1,000/month on the home, or $105.76 per month. Escrow payments pushed this figure up to $158.64 per month, with allowance for school and real estate taxes. In the 12 years that he has lived in the house, inflation has raised the escrow payments, so that his current house payments are $190.42. Obviously, this is a basic fixed cost, which could not be very easily lowered. The house could perhaps be remortgaged to raise some cash or perhaps to re-extend the period on the balance to 30 years; however, the 9% or greater rate on the new mortgage would actually raise the payments significantly. He is better off to leave this matter where it stands.

However, the house mortgage is only one of the expenses that he must face. To try to obtain some index of what the bare-bones minimum budget would be for this family of four, I spent some time in analysis of this matter with an expert on the subject, our Esteemed Treasurer. She went over a few records, and we came up with the following budget for Mr. HA and family.

BARE-BONES MONTHLY BUDGET

Home mortgage and escrow	$190.42
Water, heat and light	85.00
Telephone	12.00
Home insurance	10.00
Insurance for two cars	30.00
Groceries and soap, household exp.	180.00
Clothing for four	10.00
Maint. and gasoline, two cars, 1,000 mi/mo, 15 mpg	100.00
*Life insurance	10.00
*Health insurance	30.00
*Business entertainment	10.00
*Business telephone	5.00
*Office supplies and postage	7.00
*Miscellaneous business expense	35.00
Home maintenance	10.00
Total	$724.42

*Items marked represent new expenses due to the new business or to replace coverage formerly provided by employer.
Note: Figures referent upstate N.Y. and Midwest Mich. and Ohio; N.Y.C. area is probably higher.

From the list you may see that it is possible to cut the family budget in this case about in half. There is also some $92 included that is associated with the new business and which probably cannot be eliminated. In any event, we see that it is possible to cut the budget to just under half of the money formerly disposable. At this rate the $7,086 represents nearly 10 months of budget payments. At least initially, it also represents $29.53 in interest income from the bank at 5%.

Of course no one goes into this sort of thing with the thought of failure. With no income from the business, the thing will crash in just under 11 months. However, let us suppose that some income is available from the business, and that it grows at a constant rate from month to month such that at the end of 61 months the income is back to the level Mr. HA had achieved just before quitting. Obviously, the more initial income there is available, the longer the crash point is extended. Assuming that the growth rate is adjusted to bring the consulting income at 60 months to just $21,946, we obtain the results shown in Table 6-1, assuming starting savings of $7,086. Table 6-2 is calculated to show the initial rate that would *first* permit success under the assumed starting conditions.

Several things about Table 6-2 are noteworthy. First and foremost is the fact that Mr. HA could find it very easy to go broke, even with a business that is growing at a rate which would restore his income in five years. Furthermore, the real bind does not come near the beginning, but rather is stretched out at about two and a half years after the start. The phenomenon is rather similar to what would occur in the launching of the X-1 or one of the other rocket planes. The craft would be carried up to some alti-

Table 6-1

Initial monthly consulting	Crash month	Growth Rate per mo.	First-yr. income	Second-yr. income
$200	17	1.03757	$3,222	—
250	20	1.03372	3,897	—
300	27	1.03059	4,560	$6,233

Table 6-2

1·02895	Growth Rate M		7·		13·		19·
			391·63		464·77		551·58
			20·78		13·12		6·88
1·	Month		738·76		753·39		768·32
330·00	Income		4,661·18		2,874·51		1,443·44
29·52	Interest						
724·42	Expen.				14·		20·
6,721·10	Svgs		8·		478·23		567·55
			402·97		11·97		6·01
			19·42		755·86		770·83
2·			741·18		2,608·85		1,246·17
339·55			4,342·39				
28·00					15·		21·
726·79			9·		492·07		583·98
6,361·87			414·63		10·87		5·19
			18·09		758·34		773·35
			743·61		2,353·46		1,061·98
3·			4,031·50				
349·38					16·		22·
26·50			10·		506·32		600·88
729·17			426·64		9·80		4·42
6,008·59			16·79		760·82		775·89
			746·04		2,108·77		891·40
4·			3,728·90				
359·49					17·		23·
25·03			11·		520·98		618·28
731·55			438·99		8·78		3·71
5,661·56			15·53		763·31		778·43
			748·49		1,875·22		734·97
5·			3,434·93				
369·90					18·		24·
23·58			12·		536·06		636·18
733·95			451·70		7·81		3·06
5,321·10			14·31		765·81		780·98
			750·94		1,653·29		593·23
6·			3,150·01				
380·61			··········				··········
22·17			4,915·30	Ann. Income			6,648·60
736·35			··········				··········
4,987·53							

Table 6-2 (continued)

```
25•                    37•                 43•
654•60         31•     921•95           1,094•14
  2•47        776•85     1•10               5•41
783•53          0•35   814•88            831•02
466•77        799•05   373•41          1,567•63
              62•88

26•                    38•                 44•
673•55         32•     948•64           1,125•81
  1•94        799•34     1•55               6•53
786•10          0•26   817•54            833•74
356•16        801•87   506•06          1,866•23
              60•82

27•                    39•                 45•
693•05         33•     976•10           1,158•40
  1•48        822•49     2•10               7•77
788•67          0•25   820•22            836•47
262•02        804•29   664•04          2,195•95
              79•27

28•                    40•                 46•
713•11         34•    1,004•36           1,191•94
  1•09        846•30     2•76               9•14
791•25          0•33   822•91            839•20
184•97        806•92   848•26          2,557•83
             118•97

29•                    41•                 47•
733•75         35•    1,033•43           1,226•45
  0•77        670•80     3•53              10•65
793•84          0•49   825•60            841•95
125•65        809•57  1,059•63          2,952•98
             180•69

30•                    42•                 48•
755•00         36•    1,063•35           1,261•95
  0•52        896•01     4•41              12•30
796•44          0•75   828•30            844•71
 84•73        812•22  1,299•09          3,382•53
             265•24

          ............           ............
            9,245•62              13,073•89

          ............           ............
```

Table 6-2 (*continued*)

```
      49 •                    55 •                    61 •
  1,298 • 49             1,541 • 00             1,828 • 91
     14 • 09                28 • 12                48 • 66
    847 • 47               864 • 26               881 • 38
  3,847 • 64             7,454 • 55            12,675 • 81


      50 •                    56 •                    62 •
  1,336 • 08             1,585 • 61             1,881 • 70
     16 • 03                31 • 06                52 • 61
    850 • 25               867 • 09               884 • 26
  4,349 • 50             8,204 • 14            13,726 • 12


      51 •                    57 •
  1,374 • 76             1,631 • 52
     18 • 12                34 • 18
    853 • 03               869 • 93
  4,889 • 34             8,999 • 91


      52 •                    58 •
  1,414 • 56             1,678 • 75
     20 • 37                37 • 49
    855 • 83               872 • 78
  5,468 • 45             9,843 • 39


      53 •                    59 •
  1,455 • 51             1,727 • 35
     22 • 78                41 • 01
    858 • 63               875 • 63
  6,088 • 11            10,736 • 12


      54 •                    60 •
  1,497 • 64             1,777 • 36
     25 • 36                44 • 73
    861 • 44               878 • 50
  6,749 • 69            11,679 • 71


                    • • • • • • • • • • • •
                      18,652 • 07

                    • • • • • • • • • • • •
```

tude and dropped from a mother plane. It would first dive to pick up speed and then finally accelerate away under its own power. Obviously for every given altitude some minimum rate of acceleration would permit it to get up to climbing speed before it hit the ground. If it were taken higher initially (more than $7,086 savings) or accelerated faster (more than 1.02895/mo growth rate), it would make it more easily.

We will not get into taxes until the next chapter; however, it should be noted that some of the expenses listed are deductible business expenses; therefore Mr. HA would not have to pay tax on the entire amount. However, even from the tables you will note that he is not out of the woods. At the end of the third year, he is earning $9,245, which would mean that he might have to pay some taxes; however, you will note that he has only $265 in the bank! The performance listed is only marginally successful!

Obviously there is nothing causal involved in this simple mathematical example. The consulting business or entrepreneurship will grow because of the way that your clients respond to what you offer, not because of some simple mathematical formula. However, the model can be used to draw some basic and broad conclusions about your chances of success or failure, as follows:

1. After a careful budget paring, you should have not less than 10 months' basic budget available in savings. Anything less than this will require the attainment of an improbable rate of growth to permit success.
2. Fairly shortly after starting, on the order of six months, the business should be able to support about half of the monthly outlay. If it doesn't, the outlay has to be trimmed to fit or the whole idea abandoned.
3. At the end of the first year, if the savings are less than half of the initial amount, you are in for some very rough sledding and perhaps failure.
4. Just because the business is growing, don't assume that you cannot fail. If the initial savings or capitalization is as small as 11 months' budget, growth will have to be greater than 43% per year in order for you to survive. Greater initial capitalization will permit slower growth; however, this rate does not fall rapidly, and it implies a much longer break-even time.

As a word of amplification on the last point, let us assume that for some reason Mr. HA had an initial savings of $10,000 and was able to start the business with the same budget but a growth rate of only 1.024 per month or 32.9% per year. Suppose that he also was able to get an initial consulting income of $310/month. The operation could survive; however, he would experience a low point when the savings fell to $186 at the 42-month mark, and it would take him roughly $6\frac{1}{4}$ years to get back to an income from consulting of $21,946! At this rate it is highly unlikely that he would ever fully recover his investment in the business, considering the losses he sustained with the termination of his employment.

INVESTMENT RECOVERY

In order to obtain some reasonable assessment of the rate of projected investment recovery, we need a somewhat more sophisticated model of the economic picture than that used for the previous table. This model is presented in Table 6-3. Here, as in the previous chapter, we have applied a decaying growth rate to account for the saturating effect of the existing work. After seven years, the growth rate shrinks to 37% of the initial growth rate, and after about 21 years it has shrunk to just 2% above inflation. This factor accounts for the fact that as the practitioner gets busier and busier, it becomes progressively more difficult to fit in new clients.

For the model, the federal income tax (F.I.T.) is calculated using the model of page 76. Bank interest is taken as 5%. The change in the rate of increase in the budget is included to account for the fact that by the time some money starts to roll in, the cars, rugs, clothing and everything else around the house will be pretty well worn-out and in need of replacement, since the austerity budget required at the outset did not provide much for replacement.

An examination of the data in Table 6-3 indicates that the income is pretty well back to the 1976 level by 1982 when allowance is made for the income tax. The initial savings of $7,086 would have grown to $9,496 if left in the bank at 5%. This amount is about matched in 1982; however, this money is the capital for the business too, and is probably not completely available.

Table 6-3

```
    0.46  r₀
2,800.00  R₀
```

. . . .

Year	1,977.	
rate R_n	1.50	
C. Inc.	4,800.30	after F.I.T.
F.I.T.	0.00	
Σ Svgs	3,913.30	
	1,978.	
	1.43	
	6,480.36	
	57.38	
	1,266.01	
	1,979.	
	1.38	
	8,533.03	
	306.98	
	215.01	
	1,980.	
	1.33	
	11,038.00	
	715.37	
	1,189.77	
	1,981.	
	1.29	
	14,008.57	
	1,295.35	
	4,229.42	
	1,982.	
	1.26	
	17,417.72	
	2,059.79	
	9,691.02	
	1,983.	
	1.23	
	21,242.06	
	3,016.52	
	17,900.91	
	1,984.	
	1.20	
	25,449.92	
	4,167.29	
	29,145.75	

Rate of increase for n'th year

$$R_{n+1} = R_n(1 + R_0\epsilon^{-t_n/7} + 0.04)$$

$$t_n = \text{Years after 1977}$$

Expenses inflate from \$724.42/mo at 4% through 1979 and 9% thereafter.
Consulting income includes 5% interest on savings and deducts F.I.T.

Some of it could also have gone for improved facilities and equipment capital investment. The bank account will probably not be truly back to normal until sometime in 1983.

Had Mr. HA kept his job, he would, in the same year, be earning 6.252 times his starting salary of $5,329, or $33,317. His bank account would contain 0.34678 times this salary, or $11,553. Thus it will take him about three more years till he overtakes and passes the projected earnings at the old firm. The discrepancy in the savings is due to the more frugal habits that are pretty much required of him in the self-employed condition. This savings account is the only real buffer the self-employed person has against the vagaries of the business climate and the foul winds of chance. He has to be considerably more prepared to weather the storms, since, for example, he is likely to be ineligible to collect unemployment insurance.

This is a fairly complicated situation; hence some of the results have been presented graphically in Figures 6-1 and 6-2. Figure 6-1 provides a comparison of the salary from the job and the minimum successful consulting income. If we are to presume that the curves would cross in about 1985, we see that the area between the curves represents a loss of income, which would have to be made up in subsequent years. This disparity amounts to a gross loss of $95,755 before F.I.T. The real loss of disposable income would be closer to $60,000. Naturally a growth rate in the consulting income that is faster than the minimum required for survival would reduce this loss.

Figure 6-2 is in some respects even more revealing because it shows Mr. HA's near disaster. In this curve, the savings account is shown as a function of time for both the job and the consulting situation. The disastrous dip at the end of the third year is vividly displayed.

Obviously what has brought this crisis to pass is the fact that the income has just barely caught up with the expenditures, which have been growing at an inflation rate. By the end of the fourth year the income has passed the outgo rate comfortably, and the savings have begun to respond. After the end of the sixth year, the savings growth curve is marked with a series of question marks,

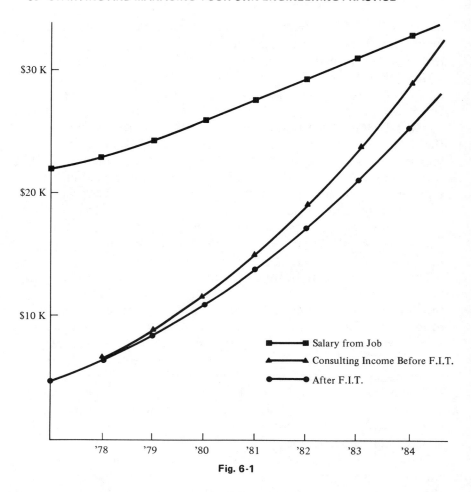

Fig. 6-1

since it seems unlikely that Mr. HA's family would adhere to what is still an austerity budget.

The difference in the growth rates of savings for the two curves is due to the austerity budget. Whereas in the job position HA was depositing 10% of disposable income (except during the college transient), in the consulting situation he deposits every nickel above the budget into savings. The small increase in growth rate of expenditures after 1979 would account for no more than replacement of worn-out items that could not be put off any longer.

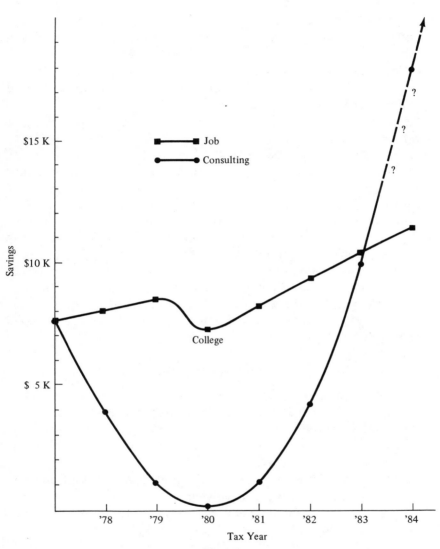

Fig. 6-2

After 1983 it would be difficult for the most frugal individuals to put off a vacation, a dinner out or some of the other comforts of life.

You will note that in the job situation the college dip is reflected in the savings curve, whereas in the consulting situation depicted it is not. The reason is that family income is below poverty level during the two years preceding college entry in this consulting situation. In all likelihood full scholarship aid would be available. Failing this, it would be necessary for Mrs. HA to obtain enough employment income to handle tuition, and so on. In this regard it is important that the two-previous-years lag in scholarship aid be kept in mind.

Naturally some additional income from outside employment of Mrs. HA or the children could save an otherwise losing situation in terms of inadequate growth rate.

The self-employed person has certain benefits in the form of tax breaks and other considerations that are not available to the person employed by someone else. For this reason, a comparison of dollar-for-dollar income is a bit misleading. The self-employed person does not have to have quite as large a reportable income in order to live as well.

To summarize, a self-employed consultant can expect to be:

1. Pretty darn near broke the third year. This is the crucial point. If he makes it past the third year his chances of survival are good.
2. Coming out of the woods about the fifth year, with income about on the same scale that he left.
3. In possession of a bank account that begins to look the way it used to at about the sixth year.
4. At about the eighth year, overtaking and passing the point where he would have been had he not become self-employed. He will begin to reap some significant net financial benefits.

It is worth reiterating that the growth rate is a highly significant factor. Unless the business grows at a rate of 43% per year or more at the outset, the chances of ever recovering the investment are slim indeed, even if the capitalization is sufficient to permit success eventually. Since a growth rate much in excess of 50% is unlikely, it can be seen that there is a very narrow "window" through which a successful conversion to self-employment can proceed. This probably accounts for the fact that a large number of self-employed people both in and outside of engineering report very similar experiences.

It is noteworthy that the analysis turns out very much the same when performed for Mr. Outstanding. His fixed expenditures are higher, and he has further to go in order to recover. He may have more to offer, but he has further to climb.

It can, of course, be argued that the timetable is based upon income alone. This is certainly not the only criterion upon which people will judge the value of self-employment. There are matters of self-determination, pride and enjoyment of one's work involved. In fact, the requirement that I get back to my old standard of living was not one of my prime criteria. If the change were to result in some net diminution of my lifetime earnings, then so be it!

On the other hand, I was to discover that the people who generally succeed at this sort of thing are fiercely competitive. They tend to succeed because they have a very strong drive to success. Now a person may be a success if he has peace of mind, happiness, a sense of accomplishment or a sense of duty fulfilled. But engineers like to measure things, and in this world of sin and sorrow, the only way to *measure* business success is money. The money may not be important of and by itself, but it tells the self-employed person that he is not the fool that a great many of his friends and relatives thought he was; that all of the pain and strain and deprivation of his family was not caused by just a foolish and irresponsible whim. There is a powerful sense of pride when the self-employed adventurer starts to regain the earning power that an employer once thought he was entitled to.

In closing, I want to repeat that there is no causal relationship implied in this analysis. If your efforts are accepted by your clients, you can succeed; if they are not, you will fail. The mathematical model presented is only a description of a typical history of what happened to a number of self-employed people who *did* succeed.

7

Tax Breaks and Benefits

If you are about to make the leap from employee to self-employed status, welcome to a bewildering and ever-changing morass of tax considerations that have never concerned you before. Chances are that as a salary earner you've been involved in a minimum of record-keeping, and have been able to prepare your tax returns yourself or get by with the help of a neighborhood, store-front tax return–preparing service. In both respects, things are probably about to change for you.

First, let's look at some of the assumptions upon which the following comments are based. It is assumed that you are operating as a sole proprietor who is truly "self employed," and that no corporation, partnership or other entity is involved in the conduct of your operation. Many, but not all, of the principles discussed apply to other business entities, but comparisons are not involved in this discussion. Second, what follows purports to relate only to federal income taxes; state and local taxes vary widely from one locale to another across the country, and cannot be considered here. Third, it should be understood that there are certain deductions to which one may be entitled under various conditions whether he is an employee or is self-employed. Included in this category are, for example, interest, certain taxes, charitable deductions, medical expenses (subject to limitations) and certain others.

Next, let's consider a few basic principles. Business expenses are deductible if they are (1) ordinary and necessary, (2) paid or incurred during the taxable year and (3) directly related to carrying on a trade, business or profession. An overall limitation of "reasonableness" is also usually imposed. "Ordinary" is generally defined as an expense that is "customary" or "commonly incurred" in the business or profession of the particular taxpayer. What might be an ordinary expense for a professional deep-sea diver might not be an ordinary expense for a consulting engineer, and vice versa. "Necessary" is usually considered to mean not absolutely essential, but appropriate and helpful. A consulting engineer might be able to get along with only an abacus, but a sophisticated computer is probably appropriate and helpful, hence "necessary." In many areas a distinction must be made between deductible business expenses and capital expenses. The latter, while not immediately deductible, are usually deductible over the useful life of the asset acquired through a depreciation deduction. Basically, the cost of machinery or equipment or other assets having a useful life in excess of one year is a capital expenditure, which cannot be fully deducted in the year of acquisition but may be recoverable over a period of years through a depreciation deduction. Further, it should be kept in mind that personal expenses, as opposed to business expenses, are not deductible (exceptions to this statement include the items listed at the end of the second paragraph above).

A strong personal recommendation to those of you about to venture into the field of the self-employed is that, at the outset, you put yourselves in the hands of a good accountant. Over the years the tax laws have been becoming increasingly complex, and they seem to become more so each time congress enacts simplifying legislation. A good accountant will understand these complexities and, in addition, will know about new rules, regulations and decisions that you cannot possibly keep up with. Remember, too, that the burden of proof is always on the taxpayer to substantiate his deductions; for this purpose maintaining accurate and adequate records is essential. Your accountant can tell you what records to keep and how to keep them, and you will not waste time amassing unnecessary materials. His fees may seem a bit high, but the odds are very good that he will save you much more than the fees you pay him. And, as a bonus, his fees themselves are deductible (as are certain attorney's fees; but remember the rule against deducting personal expenses, so do not expect a deduction, in general, when your attorney prepares your will or gets you a divorce).

Now let's take a random and noncomprehensive look at certain specific items related to income taxes with which you may find yourself concerned.

1. *Rents.* If you rent the premises from which you operate your business, rental payments are deductible. Similarly, you may deduct rentals paid for machinery and equipment, except in certain cases where your rental contract gives you the right to acquire title to the property at the end of the rental term for no or nominal consideration. In the latter case, you may be considered to have bought the equipment, and may have to look to a depreciation deduction. Items you pay for in connection with your rental premises, such as utilities and security and janitorial services, will also be deductible.

2. *Home Office Expenses.* Many new businesses start out operating in a portion of the sole proprietor's home. Certain deductions are available, but allocation of business expense and substantiation may become complicated. If you rent the home, a portion of what was previously nondeductible rent may become deductible based upon a reasonable allocation of the rent between business and nonbusiness use. If you own the house, you continue to have real estate taxes and mortgage interest payments as deductions, but appropriate portions of many more items may become deductible. These items may include insurance, utilities (more about telephones later), cleaning services, repairs and maintenance, as well as an allowance for depreciation. But you must be able to show that a specific part of the residence is set aside and used exclusively on a regular basis as a principal place where you conduct your business or meet with your customers and clients. And overall limitations on the amount of deductions may come into play to prevent total deductions from exceeding the gross income from the home-operated business.

3. *Compensation of Employees.* Subject to the limitation of reasonableness, the compensation you pay to your business employees will be deductible.

4. *Advertising, Dues and Subscriptions.* Usually, ordinary and necessary advertising expenses bearing a reasonable relation to your business activities are deductible, along with dues to professional societies and subscriptions to technical magazines and newspapers. Normally, the original cost of technical books cannot be deducted (useful life in excess of one year) but can be depreciated.

5. *Telephone Expense.* If you have a home office, the business portion of your telephone bill is deductible. If you do not have a home office but use your home telephone partly for business, you can still deduct the percentage of the telephone bill that is allocable to business use. Here the key is record-keeping so that a reasonable allocation can be made and the deduction substantiated. Some consideration might be given to installing a separate telephone, listed under the business's name, and using it exclusively for business calls.

6. *Depreciation.* This item we can leave in the capable hands of your accountant, except to say that a reasonable allowance, called depreciation, may be deducted each year for the exhaustion, wear and tear and normal obsolescence of property you use in your trade or business. There are several methods of computing this allowance, and your accountant will choose the best one for you. He will also know about an additional first-year depreciation deduction and an investment credit that may be allowable for the year in which assets are first placed in service and is in addition to the depreciation allowance.

7. *Business Travel and Cars.* In general, business travel expenses are deductible if they are reasonable and necessary expenses for transportation, food or lodging incurred while away from home. They may not be extravagant under the circumstances, and they must be directly connected with and necessary or appropriate to the business. As to cars, expenses that can be considered "commuting" are not deductible (they are "personal" expenses), and an allocation must be made between personal and business use. Alternative methods are provided for handling the computation of the business portion of the expense. If you choose the "actual expense" method, you must keep a complete, accurate and contemporaneous diary that shows in great detail the amount spent, when and where you traveled and for what business purpose, and you

must have receipts for every expenditure in excess of $25.00. Under the "automatic deduction" method, you are allowed a deduction of 15¢ for the first 15,000 business miles you drive each year, and 10¢ per mile for any excess business mileage over 15,000. But you must still keep a diary showing the business mileage traveled, when and where you traveled and for what purpose.

8. *Casualty Losses.* If you are unfortunate enough to lose business property in a casualty such as a fire or flood, you may have a deductible loss. The amount deductible is generally the adjusted basis of the property (cost less depreciation) reduced by its salvage value and any insurance recovered. If the loss is attributable to theft, you should be in a position to prove that the property was actually stolen and not merely lost or mislaid. Filing a report with the police is helpful in this connection. The amount you deduct is the lesser of the basis or fair market value of the property at the time of the theft.

9. *Entertainment.* This is usually a sticky area and subject to careful scrutiny. Lavish or extravagant expenses are disallowed. Other entertainment expenses, in addition to the ordinary and necessary expenses of the business, will usually not be allowed unless it can be shown that they were actually made to generate business and are directly related to the business. In this area record keeping is vital, and the deduction is available only where a detailed record is kept as to the amount spent and the time, place, business purpose and business relation of the person entertained. This record should be backed up with itemized receipts wherever possible.

10. *Self-employed Retirement Plan.* You may have thought that when you gave up your job as a salaried employee to become self-employed there would be no pension possibilities in your future. This is not necessarily true, because you may elect to put into effect a so-called Keogh or HR 10 plan. If your plan qualifies, you may have an annual deduction for contributions under it equal to the lesser of $7,500.00 or 15% of your earned income. In addition, Individual Retirement Plans are available to those who are not active participants in any other tax-sheltered retirement plan. Under a qualified plan of this type you may contribute and deduct each year the lesser of $1,500.00 or 15% of earned income or, if your plan includes a nonemployed spouse, $1,750.00 or 15%

of earned income annually. Under these plans the contributions may be invested in a number of different ways. In addition to the advantage of not using after-tax dollars, the increment on the investment is not taxable now. The two plans are not exactly interchangeable. The differences have mainly to do with provision for your employees. Get current professional advice before making a selection. The day of reckoning comes when you retire, but at that point a favorable tax treatment may be arranged, and you will probably be in a much lower tax bracket. In connection with these plans, your banker, broker or insurance man may be helpful, and other persons are available who are specialists in setting up a proper plan that will qualify for these benefits.

11. *Maximum Tax.* As to personal service earnings, including the fees of a professional or other independent contractor, the tax law contains a provision under which the maximum tax rate on earned income cannot exceed 50%. This may not be helpful unless you have been in business for some time, but your accountant should let you know when this 50% limitation will become beneficial to you.

It must be stressed that the foregoing discussion by no means includes all the income tax matters with which you may become concerned as a self-employed individual. Such a discussion would fill several books. Furthermore, the tax laws are, by nature, evanescent. Interpretations of what is "ordinary" and what is "necessary" vary from year to year depending upon the actions of congress, and from week to week depending upon the interpretations of the courts. You will discover that many of the large accounting firms put out a "Tax News Letter" on a monthly basis for their clients, describing changes in tax law interpretation resulting from current court cases.

Engineers going into self-employment are generally avid do-it-yourselfers. They usually feel that there are very few areas where they cannot learn to do the things required and consequently save themselves a great deal of money. You certainly can and probably should learn the art of bookkeeping for your business; however, the services of an experienced accountant in setting up the books in a convenient and advantageous manner is an invaluable help.

It is the habit of your humble servant to assign to professionals all assignments in watchmaking, brain surgery and tax accounting.

8

Office, Stationery and Style

The items to be discussed in this chapter are subject to one's personal taste. Things suitable to one person may be entirely unsatisfactory to another; and what one person considers to be attractive and distinguished, another will find to be tasteless and plebeian. One's actual requirements are also a function of the nature of his practice. For example, an architect will have an entirely different set of requirements compared to those of a mechanical engineering consultant. These requirements also vary with the modus operandi of the individual.

THE OFFICE

It is likely that anyone in the private practice of engineering will require some kind of office. There must be some place to sit down, a flat surface upon which to write, sources of light and heat, bookcases, a telephone, pencils and paper. Such accommodations could be found in a variety of places, ranging from a room at the YMCA to a corner suite atop the Trade Center in New York City. The question to be answered is what position along this wide spectrum will optimize your chances of success.

Obviously, the corner suite in a fashionable skyscraper is the most impressive location to clients; however, it is also the most expensive, and, as noted in Chapter 6, the principal early hazards in a private practice are financial. An expensive and flashy office is therefore not necessarily optimum. On the other hand, a cubbyhole office in a dingy upper hall may cost little but could well turn clients off. It is worthwhile to spend a little time considering the functions your office will serve to visitors before jumping to a conclusion about what will best suit your needs. We shall start by asking a few questions.

1. _Who is going to visit your office?_ The answer to this question is obviously a function of the nature of your practice. If you are a civil or an architectural engineer and expect to do planning and design for municipal buildings or county highways, it would be reasonable to expect that politicians, officials, prominent citizens and perhaps reporters will be visiting. If, on the other hand, you are a chemical engineer who specializes in solving problems in catalytic petroleum cracking plants, you may have no visitors at all. All of your work may be done at the clients' facilities.

Most technological consultants will fall near the lower end of this spectrum. They may have relatively few visitors every month, but will do most of their work and attend the majority of their meetings at the facilities of the client.

2. _How large will the groups be?_ The city council, officials, and so forth, could turn out to be quite a throng. On the other hand the technological consultant will generally have as visitors one or two engineers, with three or more being a rarity. Time taken to visit your office by people other than the manager and the project engineer would be considered excessive by the client.

3. _What do you do for the visitors?_ The architect or civil engineer could show large drawings, slides and, on occasion, movies to illustrate his proposed work. The mechanical engineer would also show large drawings and perhaps slides. An electrical engineer might keep his circuit diagrams smaller and would usually have fewer pictures. He is more likely however to feature a "live" demonstration of an instrument or apparatus, perhaps a computer terminal.

The point being made is, of course, that the real needs and requirements for your office are based upon your response to these questions. For the typical technological consultant, an office with seating arrangements for 20 visitors is a useless extravagance, whereas a large clear table for spreading out drawings may be a welcome and useful facility. At the start, any expenditure that is unwarranted is a hindrance rather than a help.

Another item that will affect your office space requirements is the presence (or absence) of help. Will you have an assistant or a secretary? The merits of having the help will be discussed shortly; however, if you are planning an office, the need to provide space for help should be considered.

The Rented Office

If we drop the idea of a penthouse in the Twin Towers, there still remains the possibility that a rented office in a shopping center or a professional center might be the optimum approach despite the added budget burden. The reasons for this stem principally from the businesslike image projected to visitors and the relative freedom from interference.

On the other side of the coin are a few drawbacks:

1. For the technological consultant, at the start, you will spend very little time in the office. You will be out trying to drum up business. If a full-time secretary is beyond your budget, the office will be unoccupied most of the time, and it may be very hard for people to reach you.

2. The new office will also have a new telephone, which is liable to be unrepresented in the directory for up to a year. You will have to personally inform even old friends and prospective clients of the new phone number and the office location.

3. You will probably hesitate at first to duplicate reference texts, files and other items. Following Murphy's Law, these items will always be in the office when a client finally contacts you at home, and vice versa.

4. Most of the reasonably priced office space in your community is likely to be reasonably priced because of a shabby neighborhood, difficult parking, hard-to-find location, and the like. Under these circumstances, the office rented to impress clients could have just the opposite effect.

The Home Office

Advantages. The home office has several very substantial advantages. First and foremost among them is the matter of economy. You probably already own the property, and you may already have the office. In any event, if you are at all handy with carpentry, you can probably work up an acceptable office for a rather limited expenditure in materials and time.

Another significant feature is the fact that your office expenses are a business expense and thus deductible on your federal income tax return (see Chapter 7). In this manner, you can actually have the home office turn up in the transaction as a negative expense, i.e., you will pay less tax because of the fact that you have an office in your home. This is essentially the same as some income.

Another source of savings is the fact that it may not be necessary to have another set of telephone and utility bills, and as a matter of fact portions of the existing utility bills *may* be deductible as business expense. Furthermore, you will retain the telephone number and address that your friends have for you, and you will be listed in the telephone directory for the first year— which can save you some mailing expense.

Still another advantage is the fact that, unlike the rented office, your home will be occupied much of the time by someone or other in the family, a feature that provides a substantial advantage in receiving shipments of parts and supplies. In general, you will not have extensive credit with suppliers, and they will want to ship COD or prepaid. If the postman or the UPS driver arrives at the office with a COD package, there will be someone there to hand him a check and receive the material. Also, in the matter of telephone calls, there will be someone there most of the time to field calls and make a judgment about whether they should be forwarded.

The final advantage concerns convenience. During the first few years you will be called on to be out a good deal of the time. Furthermore, you will be working very long hours. It is a considerable convenience to be able to turn off the lights in the office and not be faced with a crosstown drive when you are exhausted. Furthermore, all of your records and reference books and supplies are in one place. You can work for an hour in the evening without having to chase in to the office to get the things you need.

All of these advantages are so substantial that a number of physicians and virtually all of the outside salesmen and manufacturers' representatives in my area use home offices. When I call to order a line-printer or a reed relay, the number I reach is the home of the representative. The person I speak to is his wife. He is out making calls on different firms. When he checks in, he receives the message. This is a system which can be made to work very well and will result in considerable savings. There are, of course, some considerable drawbacks.

Disadvantages. Much of the treatment to this point has taken the view that the prospective engineering consultant is married. Statistically, this is a fair bet for someone with the experience necessary to have a reasonable chance for success, but it is obviously not a causal condition. An unmarried engineer could just as easily decide to go into private practice and would have certain advantages from an economic standpoint. However, the chances that an unmarried engineer would also own a home in which to have a home office are smaller yet. I therefore beg the indulgence of the unmarried reader. Most of the disadvantages as well as most of the advantages of the home office are associated with the state of wedded bliss.

Probably the largest single group of disadvantages associated with the home office stems from the fact that there are other people living in the structure. If your youngest child is less than 40 years of age, there will almost certainly be interruptions, which can range from an inoperative music box in Big Bear to an automobile that will not start. It can be a bit difficult to enforce; however, a good rule to follow is: THE FACT THAT YOU ARE ON THE PREMISES DURING WORKING HOURS DOES NOT MEAN THAT YOU ARE AVAILABLE FOR FAMILY MATTERS. (In passing I might note that I have seen the careers of persons in the employ of a large corporation hurt by the failure of families to observe this form of courtesy.)

Another drawback to the home office stems from the fact that the other people living there originate and receive telephone calls.

This can be disruptive of the work flow if you answer the telephone in the office and the call is for one of the family. Also, it is annoying to attempt to make a business call and find that the facility has been preempted.

It would seem that this problem would be easily solved with the installation of a separate telephone line. Unfortunately, it does not work out quite so easily. If you get the new number for the business, you have the problems previously discussed under "The Rented Office." If on the other hand you elect to have the family use the new number, you will continue to receive 99% of the family calls. The addition of a "children's telephone" listing to the telephone directory may be of assistance in this matter. However, it will control only the calls made by people who look the number up in the telephone directory. Those who use their personal phone index will continue to call the old number. It takes a substantial period of time for this effect to disappear. In addition, the family will not answer the office phone; therefore, one of the advantages previously discussed is lost.

From a standpoint of cost-effectiveness and workability, the best solution rests with a little self-discipline:

MINIMIZE PERSONAL CALLS DURING WORKING HOURS.
MINIMIZE LENGTH OF CALLS.
ACCEPT NO CALLING-COMMITTEE ASSIGNMENTS.

It has been pointed out to me by a professional woman that this matter of telephone discipline is particularly troublesome to self-employed women with a home office. Non-business calls from chatty neighbors, the PTA, the league of women voters, etc., can occupy a considerable amount of time and must be dealt with diplomatically. One does not want to alienate the callers but the time must be kept in hand. One suggestion here is the fact that it is relatively simple to rig a surplus telephone bell to ring with a push-button. When the unwanted call is identified one simply pushes the button and says "excuse me but there is a call on the other line." After a few tries, the party on the other end will know that you really are busy.

Another topic demands a bit of attention. Depending upon the design of the house, it may be possible to have the office entrance completely separate from the living quarters so that visitors need not enter through the living area. This is the best situation. If it is not feasible, then strict discipline must be enforced regarding the neatness of areas through which the visitors will pass. If a home office is not terribly impressive to begin with, the presence of piles of schoolbooks, dirty ashtrays, laundry and other family impedimenta on the way to the office will make it still less so.

THE HOME OFFICE AND ITS APPROACH SHOULD BE KEPT NEATER AND MORE ORDERLY THAN A REGULAR BUSINESS OFFICE. The matter of neatness and order extends to the conduct of the other residents. When a visitor is present, he should not be exposed to audible pollution from the Rolling Stones or the Led Zeppelin. Fights, quarrels, and physical combat among the siblings should be held to an inaudible level. *His* office is probably not that well regulated, but yours should be!

It is also worthy of note that the residential area in which Mr. Outstanding and Mr. HA live is probably zoned against any commercial or business ventures. Despite this, the physician and the real estate agent have offices in their homes, as do most of the manufacturers' representatives and the outside salesmen. Obviously one cannot put up a sign (although the physician does), and an effort should be made to keep down the number of cars parked in the driveway and the street. In general, when action is taken on such zoning questions, it is because people have been operating a sizable business in their garage and have had a number of employees parking cars in the driveway and the street. The technological consultant can usually operate his home office without any sign of a business to the neighbors. The occasional visitor is no more apparent than are the visitors to the other homes. THE PRESENCE OF THE HOME OFFICE SHOULD BE KEPT UNOBTRUSIVE AND UNOBJECTIONABLE TO THE NEIGHBORS.

Prestige and the Home Office. As noted earlier, one of the objections to the home office is lack of prestige; people tend to be more impressed with a fancy business office than they are with a home office. This effect is accentuated by the private practice syndrome. A lot of your friends, neighbors and associates will be firmly convinced that you are simply out of work. The engineer starting his own practice should be prepared for this reaction and equipped to shrug it off. A fashionable office would reduce this response slightly at a considerable financial outlay, which would in turn act counter to your chances of success. IF THE RENTAL OF AN OFFICE IS NOT REQUIRED BY CLIENT RESPONSE, DON'T DO IT!

THE LETTERHEAD

The establishment of the credibility of your new venture is probably least expensively done by means of your letterhead and brochure. We shall consider the letterhead first.

Compared to your office, which will be seen by only a minority of your clients, your letterhead will be seen by all of them, and by an even larger group of people who are potential clients. For this reason some time and care should be given to the preparation of a top-quality letterhead.

At the start of your practice, you will be concerned with informing the largest possible number of former associates and other potential clients of your entrance into the business. For local people, this can be done by physically visiting their facility or calling them on the telephone. However, for most of the out-of-town people, the personal call is not very cost-effective. The most effective way to make those contacts is by mail.

The by-mail contact has the advantage that something tangible is left in the prospective client's files. He probably does not have something for you at the moment you walk through the door; however, at some later date something may materialize, and he will be able to reach you by consulting the file. It is important that you provide him with some first-class printed matter that catches his attention in the first place and sticks in his memory thereafter.

This is a low-yield process at the start, and you will send out a good deal of mail with very little affirmative response. It is an interesting phenomenon that there is a tendency for business people to write answers to these things even when there is nothing for you. It is probably done out of politeness. You can expect that this initial mailing will receive a 40 to 60% response if the letterhead and the material sent are catchy and to the point. Most of these responses will be negative; however, you can expect that eventually some business will result from a number of these initial contacts. This is a necessary part of filling the pipeline so that an eventual smooth flow will result. At the time that you are doing it, the effort may seem a little fruitless because of the long time constant for a result. However, it is a necessary part of the business and should not be neglected.

Your stationery should have some distinguishing feature. As far as the communications aspect is concerned, plain white typing paper would serve to convey information. However, you should be attempting to convey something more. In the advertising business, this extra message is referred to as product identification. What you would really like is to have people instantly recognize your communications. If someone remembers vaguely that a person called a year ago and discussed doing something that he requires now, he will probably not remember the name and will have a problem finding the letter or brochure in the file. A distinguished logotype or letterhead can make the difference between his finding your material and its being overlooked. A logo that caught his eye the first time may be the only clue he now has to finding the material.

Preferably, the letterhead should convey some information about what you are offering. The examples shown in Figure 8-1 are intended to convey a picture of the offerer's services and at the same time give a memorable graphic image. The top one shows Kilroy (of World War II fame) being shaken by a seismic test blast. The center one has the descriptive phrase shaped to represent symbols for a hydraulic steering mechanism. In the bottom one the initials of the offerer coincide with the initials used to describe the Transverse ElectroMagnetic mode by which a coaxial line operates. The name of the offerer is emphasized by the size of the letters.

The matter of taste enters strongly into the selection of a logo. I am certain that a significant fraction of the readers will find the attempt at humor in the samples to be plebeian and perhaps unprofessional. Still others might not object to it but would view

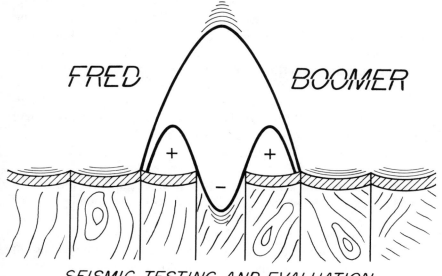

SEISMIC TESTING AND EVALUATION

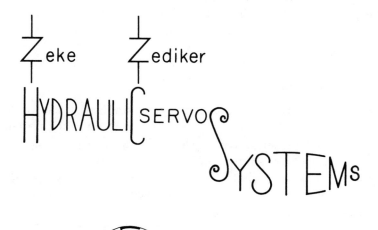

Thomas E. Mitchel
RF DEVELOPMENT

Fig. 8.1.

the effort as a little bit too "hard sell" for a professional offering. On the other hand, George Washington Duke, founder of the American Tobacco Company (Lucky Strike, etc.) was known to spit upon the conference table at advertising meetings and comment, "I just did something disgusting and you will always remember me." I feel that it is the wisest course to stay closer to the conservative end of the spectrum; however, if your own personality warrants it, a little bit of flamboyance in your stationery will not hurt. Furthermore, the offering is more likely to be remembered than a chaste letterhead with only your name and function printed upon it.

People inclined to attempt the establishment and operation of a private engineering practice are inclined to be independent and a bit flamboyant. The step itself flies in the face of established practice. If it fits your personality, let your stationery be distinctive.

A final word about design: I personally do not favor the use of abstract, arty blobs and shapes in a logo because they may convey nothing to the viewer and are, I feel, more quickly forgotten than is a representative shape that cleverly conveys an idea.

Obviously, care should be taken to have the written matter issuing from your office appear businesslike. Remember that care should also be taken to avoid anything that could be offensive.

THE FIRM NAME

Another topic that deserves consideration is the use of a firm name. In most states it is required that one register the name of the firm and obtain a DBA (*Doing Business As* . . .) certificate. This is a relatively simple and inexpensive procedure. The usual requirement is only that the name not be registered to someone else. Thereafter, you can open bank accounts and advertise your services as, for example:

ACE ELECTRONIC EXTERMINATORS
computer software debugging

Alternatively, you can simply do business under your own name. This approach has certain advantages:

1. If you have succeeded at some of the preparatory steps described earlier, quite a few people around the industry are fa-

miliar with your name. Your name is probably the major asset of the firm. If you do business under an assumed firm name, this asset will not be fully used.

2. In the matter of finances, you will find it very easy to open an account in the name of the company; however, you will find it rather complicated and time-consuming to establish credit for the Ace Electronic Exterminators. The bank has no credit history on the firm, and its assets are minimal. However, by the time you are ready to start this venture, your personal credit is probably well established, and you will have little difficulty obtaining the necessary credit. You can easily purchase air tickets or instruments on personal loans. Surprisingly, there is little if any difference in the interest rates, with the advantage, if any, going to the personal loan.

On the other hand, there are some disadvantages:

1. You will have perpetual problems with secretaries who ask: "And what was the firm name, sir?" whereupon you must explain that you are a consulting engineer. . . . This does not seem to make much difference to the clients, but it does seem to bother secretaries and order clerks at supply houses.
2. A similar difficulty may arise when you are filling out forms for free magazines and forms to put you on the mailing list for catalogs, and so forth.

A logical solution to this is to answer these with the name Fred Boomer Associates or to use the descriptive title on your letterhead, Hydraulic Servo Systems, to fill the slot.

The cost of having stationery printed with the logo will probably be on the order of $200; however, the investment is well worth it. In general you will want $8\frac{1}{2}$ by 11 letter stock, and business envelopes with the logo. It is also worthwhile to have some of the letterhead stock sheared to $8\frac{1}{2}$ by $7\frac{1}{3}$, with a line ruled across the center of the sheet. This size may be used for invoices and purchase orders by typing the appropriate designation above the line. It is also helpful to have sheets with the logo of the smaller size used on the envelopes printed two-thirds of the way down the sheet. These sheets are very useful for reports and lend a professional touch to the front cover. The question of business cards will be discussed along with the brochure.

THE BROCHURE

When you call on or write to prospective clients, it is always good to have something to leave with them. It can take the form of a business card or a brochure. I am personally convinced that the brochure is about ten thousand times as effective in this matter as a regular business card. A small box in the front of my desk drawer contains no less than three gross of business cards. Once in a while I remember that sometime ago some salesman was in with something that I might be able to use now. Then there begins a search for the card of the fellow who carried the item. I am always impressed by the fact that I cannot even identify about two-thirds of the cards in the pile or remember why I saved them, and they go in the basket.

On the other side of the coin, I am prone to keep brochures for instruments, parts, testing services, and so on, that impress me as being potentially useful. I may or may not remember the name of either the representative or the product, but I often have a visual image of the brochure. Furthermore, when I find the brochure it will say something other than FRED FOTZENGARGLE—Universal Dual Encabulators. There is usually a description of the product, and sometimes a price. I can determine something about whether the product has some promise of suitability before I make the call, and can ask a few intelligent questions of the representative. Because of this advantage in utility I feel the brochure has something to offer that is worth the sizable difference in price compared to a business card.

Another significant advantage of a brochure concerns the second-hand conveyance of information regarding your services. Often you will find that a firm in need of a consultant will assign the task of interviewing one or more consultants to one of its managers. This man probably will not have the authority to retain your services entirely on his own recognizance. After an hour or two spent talking to you and perhaps other candidates, he will be called upon to render an appraisal before his management. Unless the manager has an extraordinary memory, he will be asked questions in the appraisal conference whose answers he cannot remember. A well-written brochure will help him to select your services and justify the choice.

What Should the Brochure Contain?

A well-written brochure can be prepared on a single $8\frac{1}{2}$-by-11 sheet folded in three parts so that it will fit directly into business envelopes. The front page should carry your logo and some message to catch the attention of the reader. In preparing this document, it is probably a good idea to pay the difference for a two-color or a three-color printing job of good quality. In this matter, you might be well advised to obtain the advice of a professional artist or ad-layout man. In recent years several best-selling paperbacks have come out with two or three different covers with different color schemes and slight differences in the cover layout. Frequently one of them will outsell the other two by margins as large as three to one at the retail level. You may not be able to judge a book by its cover, but apparently you can sometimes *sell* one by its cover.

The message on the front cover should be short and forceful. It should give the reader a broad picture of the services you are offering and some reason to believe that he might profit from their use. The message on the front page should also carry your name, address and telephone number so that the reader instantly knows who is doing the advertising.

The back page should carry a picture and a succinct résumé. If you have written textbooks in your field, earned honors, published papers and received patents, this is the spot for that information. Educational honors and degrees also belong here. Most people pick up a brochure folded in three and read the front page first. They then read the back page before opening the paper. These two surfaces set the tone for the remainder of the document.

The next portion to be read is usually the back side of the front page. This is a good location for a list of clients. If you have served the Bell Telephone Laboratories or the Shell Oil Company, this is the place to let it be known.

The sheet opposite this is a good place for the caption HOW WE WORK. Beneath this caption should be a brief statement of policies with regard to patent rights and job commitment, and a brief listing of the type of services you are prepared to offer. You might indicate that you are prepared to do independent studies, work with the client staff, build breadboards and prototypes, develop unique apparatus, and so on.

When the HOW WE WORK sheet is flipped open, the remainder of the surface should be divided into WHAT WE DO, with a description of things that you have done, and OUR FACILITIES, with a description of the facilities available for performing some of the services described.

The organization described proceeds from the general to the more detailed and ticks off most of the questions that will be asked in the order in which they will be asked. For the services of someone other than an electrical engineering consultant, this organization may not be optimum; however, it seems like a fairly good starting point. Each of the separate surfaces should stand alone and have a forceful conclusion at the bottom line.

Even more than in the case of the logo, it can be argued that the approach of using a brochure organized as suggested tends more toward hard-sell merchandising than the understated shingle of the lawyer or doctor. However, it should be noted that your clientele is vastly different from theirs, and is much more accustomed to a businesslike hard sell. It is noteworthy that in this day and age firms offering such traditionally conservative services as banking and brokerage have gone over to a hard-sell appeal to the public—"Merrill Lynch is bullish on America," and the like.

Establishing your own engineering practice is a tough row to hoe, and you can use all the help you can get from marketing techniques. My own experience has been that clients are not offended by a brochure, and I feel that it has proved to be a valuable business aid.

9

Equipment and Personnel

Along with your office you are going to need certain equipment and perhaps personnel. Probably the most obvious requirements are for a typewriter and a telephone. These days it is mandatory that engineering reports be typewritten. In addition you will find that invoices, purchase orders and letters also must be typewritten. Of course, having a typewriter without someone to type is a bit futile, so let's begin by talking about the pros and cons of having a secretary.

THE SECRETARY

Before they left the giant corporation, Mr. Outstanding and Mr. HA both had private secretaries. She (or perhaps he in the case of Mr. Outstanding) would arrange appointments, answer the telephone, make airplane and motel reservations, file material, answer letters and prepare speeches and reports. At a high-enough level in the company, the secretary does not type; but let us presume that she did for our discussion here. In any event, someone did the typing.

For the person establishing his own practice in engineering, there is a great benefit in self-sufficiency. If the financial situation is at all like that described for Mr. HA in Chapter 6, the budget simply will not tolerate the salary, unemployment and Social Security benefits required to maintain a full-time secretary. However, let us suppose that somehow Mr. Outstanding does have the resources to back this service. How would it work out?

As soon as the letterhead stock and the brochures arrived from the printer, there would be a rash of letters to be sent out with the brochures. They would inform friends and associates of your new business. This initial mailing would essentially amount to a series of brief form letters to accompany the brochure. Some might contain a personal note; however, they would be pretty much the same. There would be a certain number of responses to them; however, most would be of such a nature that they would require no response. There would also be a brief stir of organization of the files and reference books in the bookcase.

After this initial activity, things will generally dip down to a few letters and a few calls a week. If the secretary is worth her salt, she will be bored to death and will probably quit. Mr. Outstanding will be out on the road trying to drum up some business, and the office will be terribly quiet. Except in the case of a new architectural or civil engineering firm, the retention of a full-time secretary is probably not warranted economically. Furthermore, a full-time secretary is probably financially out of reach for any firm not started with some form of outside funding.

It should be noted, however, that the secretarial functions remain even if there is not a great deal for the secretary to do. The letters and the phone calls must still be taken care of. How are they to be handled?

THE SECRETARIAL SERVICE

In nearly all business centers there are some very capable and relatively economical secretarial service businesses. For a nominal sum the persons who operate these services will turn out a beautiful-looking engineering report. For multiple copies they will turn the extras out on a dry paper copier kept on the premises and better-maintained than the one Mr. HA used at the corporation.

For an additional sum or monthly fee, it is possible to have matters arranged so that your office phone will ring in the service office and the person who answers there will say "Ace Electronic

Exterminators" or "Fred Boomer Associates," as the case may be. The line on which the call came in is identified for her.

These firms also do a relatively remarkable job of taking dictation over the phone and having the typed copy ready for you the next morning.

The secretarial and answering service is the high-class way to go if you are in a line where you expect to receive a lot of calls or write a lot of letters or reports. Even the occasional report will benefit from their services. Unfortunately, a certain amount of the writing of the consulting engineer will contain technical jargon that is unfamiliar to the person on the other end of the telephone. Translated into shorthand and then typed, you may discover that an LSI chip became "Elsie" or "L-C" or "Lawrence, swing, India" if you start to get phonetic. When you were with the XYZ corporation and you got a new secretary fresh from business school, you could correct her and expect that in a year or so she would learn your lovely penmanship and your vocabulary. With the secretarial service, you may not work with the same person often enough for this to happen. You will generally get better results if you send in rough-typed copy and have them turn out the finished work. If it is clear enough to read, they will generally do an excellent job with it.

THE NEIGHBOR-STENO

If you keep your ears open and ask around, you will frequently find a neighbor who would like to pick up some outside income doing typing. Frequently this is a person whose children are in school. She may enjoy a bit of spare-time income when she is free to do the work at her convenience. This can be a little less expensive than the secretarial service. The main difference is that you get the work at her convenience, not necessarily yours.

THE ANSWERPHONE MACHINE*

These gadgets are hated by nearly everyone. They squawk and screech and are constantly filled with messages from the telephone

*See Isaiah, Ch. 50, v. 2.

company telling you that the phone is off the hook. They are mostly ill-designed from a human factors point of view. For example, when the machine has been unattended for several days, there is a sizable amount of tape used up but no indication of how many calls were covered. You rewind the thing and do not know how many calls to listen for. The machine has no mechanism for bulk-erasing; therefore you may go into calls from previous weeks unless you painstakingly erased all previous calls by recording silence over them.

In addition, friends of your spouse and children will be seized by an immediate paralysis of the vocal cords when confronted by your recorded voice, and you will be treated to 30 seconds of gasping at the far end. Storm window and swimming pool salesmen, on the other hand, will immediately hang up at the beginning of your 30-second recorded message. The confounded machine will not recognize a dial tone and hang up; therefore it will trip the phone company's "off-the-hook-dead-line" circuitry, which in turn triggers the "I'm sorry but your telephone is temporarily out of order" message, which is followed by a screech to attract your attention. It seems sort of obscene to have one tape recorder talk to another over the telephone lines.

Despite all of these annoyances, the answering machine is the most cost-effective way of handling telephone calls while you are out of the office. Surprisingly, the clients do leave a message and you do get it. By the time you get around to buying one, you may be able to get bulk erase, an incoming call counter and a system to recognize the dial tone and hang up on a dead line.

I would not recommend one of the gadgets that will play the messages back to you over the phone unless you really want to pay long distance rates to listen to a series of squawks and telephone-off-the-hook messages.

THE COPY MACHINE

A great asset around the office is a small, portable dry copy machine. Some of these machines, which use treated paper, can be purchased for under $100. With a little experience you will discover that you can turn out fairly good copies with one of them.

The copies are a little more expensive than copies made on a large plain-paper copier, and the machine is considerably slower; however, it is quite advantageous to have the machine at your fingertips and not have to chase out to the bank or library to get a single copy.

For one thing, typing is much easier using a correction tape or self-correcting ribbon provided that there is no carbon copy to be corrected as well. The copies can be made with the copier. Also, it is much better to work with copies rather than originals of circuit diagrams or parts sketches. Again the copier fills in.

Frequently a client will call requesting additional copies of sketches, graphs and artwork included in a report. You can zip them off on the copier and then drop them in the mail. Computer runouts represent another common use of the copier. They are really much better copied than transcribed because of the possibility of transcription errors.

All in all, the portable copier can be well worth the investment.

THE COMPUTER

For most technological consultants, the use of a programmable computer is nearly a must. A few years ago one would take a look at a particular problem, make a few worst-case guesses and a few simplifying assumptions and slide a couple of answers out on the slip-stick. Those days are gone forever! Clients have learned that they can easily expect to have rigorously correct solutions to whole families of problems printed out in tabular form, and with higher levels of precision than was ever possible with the graphical solutions that used to suffice for complex manipulations.

In this regard, there is a great advantage to having a printing-type machine which gives hard-copy answers for scientific problems. For most scientific problems one of the desk-top scientific models is ideal for this kind of work. The programming is very simple, and the presence of the printed output makes possible the error-free transfer of the data by the simple expedient of copying the output tape on a copier. If the computations and the answer lists are quite lengthy, the probability of a transcription error is substantial.

Also, in a great many practical engineering situations, there are not enough equations to fit all the unknowns. One of the great advantages of a programmable hard-copy computer is that one can construct multidimensional parametric solutions whereby one fixes one of the variables and runs through a series of solutions for the other variable. The arbitrarily fixed variable may then be incremented, and a second and third tabular solution run off. The machine is tireless, and the only cost is electricity and paper. In this manner one can rigorously handle solutions for problems that would have been prohibitively time-consuming and expensive only a few years ago. Unfortunately, the same type of solution could be performed on a pocket programmable; however, it would prove quite tedious, and the results would be prone to transcription error after only a short span.

Along this line, the availability of a plotter to go along with the computer would be of considerable benefit. Since I have not been able to afford one of these myself, I will not recommend it at this point!

For computer software consultants, the scientific programmable is probably not the correct choice, and a completely different type of computer facility might be optimum.

LAB EQUIPMENT

It is difficult to say anything meaningful about the subject of lab equipment that would cover the full spectrum of engineering. The computer software man would have entirely different requirements from those of the hydraulic servo engineer. It should be noted, however, that each probably should have some equipment available to permit the testing and checkout of at least some of the smaller items which will be encountered. Engineering is a practical science, and the ideas and proposals should be tested with hardware whenever it is humanly possible.

In keeping with the latter point, it is possible to obtain test equipment that is too expensive to maintain for occasional use from one of two sources: the client and the various equipment rental agencies. In the case of rental, the equipment can usually be obtained for a period of six weeks to three months for a fraction on the order of 10% of the purchase price per month. This seems pretty steep; however, it is a direct business expense that uses none or little of your capital and should be 100% deductible from income. A lease/purchase arrangement is usually also available, over a period of one to five years. In this case one makes fixed payments, as on an automobile, for leasing the equipment; and the item is still probably an expense item, but you should check this with your accountant.

Sometimes, if the client has the equipment and it is available, you may be allowed to borrow it for the duration of the task. This is, of course, the least expensive procedure of all. On the other hand, if the client does not have the equipment and will require it at the completion of your task, he can on rare occasion be talked into buying it and having it delivered to your lab. Don't hold your breath for that to happen too often!

FINANCING

In the normal course of operating your business, you will at times find it necessary to borrow money or finance the purchase of materials and instruments. Also, if you are going to do any significant amount of traveling, you will find that the financing of airplane tickets will often be required. It may sometimes be worked out to your advantage.

Let us suppose that you have to make a trip to Europe. If the trip requires visits to several countries at the height of the tourist season and the schedule precludes the use of a tour ticket, the fare can easily run to $1,000. If the ticket is charged to one of the various credit card systems, there is a good chance that the card will not be billed to you for 30 to 60 days. If the client has made partial or full payment by that time, you will simply pay the amount in full without carrying charge. You will be ahead by the amount you would have lost on interest by withdrawing the money from savings. If the client has not paid before the bill comes in, you will simply pay the minimum payment on the account until he does pay.

Similarly, suppose that you decide to lease the instruments described in the previous section. The lease fee is calculated as if you were buying the instrument on a time payment plan.

While I do not recommend that the consulting engineer go about contracting for fixed overhead payments willy-nilly, there are certain advantages to handling the large transactions in this manner:

1. The interest on a loan is a direct business expense and is fully deductible for tax purposes. This tends to reduce the actual cost of such financing.
2. The cash in your savings remains available. In addition, it earns interest, which helps defray the cost of the financing. The availability of reserve cash is not to be underestimated.

Let us amplify the latter point briefly. Suppose that you have withdrawn cash from your savings account for a business trip and to purchase items for use on a given project, leaving your savings depleted. You have every expectation of being able to replenish the savings at the completion of the job. However, suppose some other major expense item comes up, such as an inescapable requirement for a new roof on the house or a new car.

At this point you could find it difficult to obtain a regular loan. In the eyes of the banker you do not have a regular job, your savings have been recently depleted, and you probably have little that would be of interest to him as collateral. If, on the other hand, your savings account contains several thousand dollars, a home improvement loan or an auto loan could proceed smoothly; and if not, you still have the savings to fall back on. As silly as it may seem, it sometimes makes good sense to borrow while you have money in savings.

Naturally, one of the things you will be interested in is the amount of fixed payment you will have to incur to carry a given debt. Figure 9-1 illustrates the monthly payment required for a loan of $1,000 versus the loan period. This problem is solved as a differential equation with the assumption that you will be paying the money back at the smooth rate shown on the curve. This is always a few cents per month less than a real finance charge, since you do not repay the loan smoothly but rather in monthly

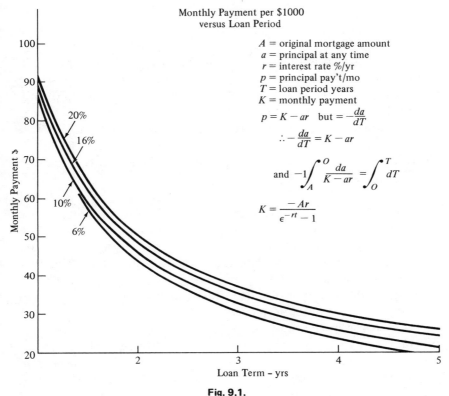

Monthly Payment per $1000
versus Loan Period

A = original mortgage amount
a = principal at any time
r = interest rate %/yr
p = principal pay't/mo
T = loan period years
K = monthly payment

$$p = K - ar \quad \text{but} = -\frac{da}{dT}$$

$$\therefore -\frac{da}{dT} = K - ar$$

$$\text{and } -1\int_A^O \frac{da}{K - ar} = \int_O^T dT$$

$$K = \frac{-Ar}{\epsilon^{-rt} - 1}$$

Fig. 9.1.

quanta (or lumps) you therefore spend most of the time owing a bit more than the differential equation would indicate. For example, the differential equation solution for a $1,000 loan over a five-year term at 18% compounded instantaneously, yields an answer of $25.28 for the monthly payment; whereas the monthly summation payment solution gives 59 payments of $25.39 and a last payment of $25.07. The difference can be seen to be negligible.

Surprisingly, the curve reveals that the difference in payments is not much affected by the interest rate for very short-term loans. This makes a bit more sense if one realizes that if the period of the loan were zero, the payment would be $1,000 regardless of the interest rate. The loan period has to extend to three years or more before the difference in monthly payments becomes more than a few dollars.

Also of interest is the question of what the loan will cost you in total dollars. Tables 9-1, 9-2, and 9-3 were adjusted to give the summation of interest payments and the balance due on a monthly report basis for a $1,000 loan at three rates of interest. If we examine Table 9-1, we see that after the ninth month at 18% interest we have paid $107.71 in interest and the balance due is $533.69. In other words, half way through the contract, less than half of the original amount has been paid off.

At the time of this writing, most credit card finance charges run around 18% and lie within the range covered by these tables. They are generally adjusted to a period of 18 months. Naturally, when you charge an air ticket or some other expense to an account, you will want to include some allowance for the finance charges that you will be paying. If you were paying 18% and the client reimbursed you three months after the charge came through, you would have spent $42.79 on carrying charges.

Table 9-4 is in the same format except that the period has been selected as five years to be representative of five-year financing of an instrument. At the end of the five-year period, with an 18% interest rate we find that we have paid $532.72 in interest. Our $1k instrument has become a $1,523.40 instrument. The thousand dollars that we left in the bank at 5.33% has earned $296.67 in interest during the same period. In addition, there is a tax credit, which would amount to about $106, because we deducted the interest payment from our income. Therefore, the real cost of the financing is about $426.16 - 296.67 = $129.51.

I am sure that some would argue with this viewpoint; however, the fact remains that we had the use of the $1,000 for the five years, and that we did not pay tax on the interest charged. Of course, we did pay tax on the interest earned by the $1k. If this tax is considered, the bill for financing rises to about $188.

The reason for discussion of credit card and personal loan types of financing is the fact that the more normal type of business financing is nearly unavailable to the engineer in private practice. If

Table 9-1.

0 • 2 0 Interest Rate
1 • 5 0 Period—yr
6 4 • 7 3 $ Monthly Payment

• • • • • • • • • • •

1 • 0 0 Mo.		1 0 • 0 0
1 6 • 6 6 $ Int.		1 2 8 • 9 3
9 5 1 • 8 8 Bal. Due		4 8 1 • 1 3

• • • • • • • • • • •   • • • • • • • • • • •

2 • 0 0   1 1 • 0 0
3 2 • 5 3   1 3 6 • 9 4
9 0 2 • 9 7   4 2 4 • 3 6
• • • • • • • • • • •   • • • • • • • • • • •

3 • 0 0   1 2 • 0 0
4 7 • 5 8   1 4 4 • 0 2
8 5 3 • 2 4   3 6 6 • 6 6
• • • • • • • • • • •   • • • • • • • • • • •

4 • 0 0   1 3 • 0 0
6 1 • 8 0   1 5 0 • 1 3
8 0 2 • 6 8   3 0 7 • 9 9
• • • • • • • • • • •   • • • • • • • • • • •

5 • 0 0   1 4 • 0 0
7 5 • 1 7   1 5 5 • 2 6
7 5 1 • 2 7   2 4 8 • 3 4
• • • • • • • • • • •   • • • • • • • • • • •

6 • 0 0   1 5 • 0 0
8 7 • 7 0   1 5 9 • 4 0
6 9 9 • 0 2   1 8 7 • 7 0
• • • • • • • • • • •   • • • • • • • • • • •

7 • 0 0   1 6 • 0 0
9 9 • 3 5   1 6 2 • 5 3
6 4 5 • 8 9   1 2 6 • 0 5
• • • • • • • • • • •   • • • • • • • • • • •

8 • 0 0   1 7 • 0 0
1 1 0 • 1 1   1 6 4 • 6 3
5 9 1 • 8 7   6 3 • 3 7
• • • • • • • • • • •   • • • • • • • • • • •

9 • 0 0   1 8 • 0 0
1 1 9 • 9 8   1 6 5 • 6 9
5 3 6 • 9 6   − 0 • 3 4
• • • • • • • • • • •   • • • • • • • • • • •

Table 9-2.

```
0•18  Interest Rate
1•50  Period—yr
63•73  $ Monthly Payment
```

• • • • • • • • • • •

```
    1•00  Mo.                10•00
   15•00  Σ $ Int.          115•72
  951•22  Bal. Due          477•92
• • • • • • • • • •       • • • • • • • • • •

    2•00                     11•00
   29•26                    122•89
  901•70                    421•31
• • • • • • • • • •       • • • • • • • • • •

    3•00                     12•00
   42•79                    129•20
  851•45                    363•84
• • • • • • • • • •       • • • • • • • • • •

    4•00                     13•00
   55•56                    134•66
  800•44                    305•52
• • • • • • • • • •       • • • • • • • • • •

    5•00                     14•00
   67•57                    139•25
  748•67                    246•33
• • • • • • • • • •       • • • • • • • • • •

    6•00                     15•00
   78•80                    142•94
  696•12                    186•24
• • • • • • • • • •       • • • • • • • • • •

    7•00                     16•00
   89•24                    145•73
  642•78                    125•25
• • • • • • • • • •       • • • • • • • • • •

    8•00                     17•00
   98•88                    147•61
  588•64                     63•35
• • • • • • • • • •       • • • • • • • • • •

    9•00                     18•00
  107•71                    148•56
  533•69                      0•52
• • • • • • • • • •       • • • • • • • • • •
```

Table 9-3.

0 • 1 6 Interest Rate
1 • 5 0 Period—yr
6 2 • 8 3 $ Monthly Payment

• • • • • • • • • • •

1 • 0 0 Mo.
13 • 3 3 Σ $ Int.
9 5 0 • 5 0 Bal. Due
• • • • • • • • • • •

1 0 • 0 0
102 • 5 5
4 7 4 • 2 5
• • • • • • • • • • •

2 • 0 0
26 • 0 0
9 0 0 • 3 4
• • • • • • • • • • •

11 • 0 0
108 • 6 7
4 1 7 • 7 4
• • • • • • • • • • •

3 • 0 0
38 • 0 1
8 4 9 • 5 2
• • • • • • • • • • •

12 • 0 0
114 • 4 4
3 6 0 • 4 8
• • • • • • • • • • •

4 • 0 0
49 • 3 3
7 9 8 • 0 1
• • • • • • • • • • •

13 • 0 0
119 • 2 5
3 0 2 • 4 6
• • • • • • • • • • •

5 • 0 0
59 • 9 7
7 4 5 • 8 2
• • • • • • • • • • •

14 • 0 0
123 • 2 8
2 4 3 • 6 6
• • • • • • • • • • •

6 • 0 0
69 • 9 2
6 9 2 • 9 4
• • • • • • • • • • •

15 • 0 0
126 • 5 3
1 8 4 • 0 8
• • • • • • • • • • •

7 • 0 0
79 • 16
6 3 9 • 3 5
• • • • • • • • • • •

16 • 0 0
128 • 9 9
1 2 3 • 7 1
• • • • • • • • • • •

8 • 0 0
87 • 6 8
5 8 5 • 0 4
• • • • • • • • • • •

17 • 0 0
130 • 6 3
6 2 • 5 2
• • • • • • • • • • •

9 • 0 0
95 • 4 8
5 3 0 • 0 1
• • • • • • • • • • •

18 • 0 0
131 • 4 7
0 • 5 3
• • • • • • • • • • •

Table 9-4. Finance Charges on a Five-year Loan of $1000.

0•20 Interest Rate	0•18	0•16
5•00 Period—yr	5•00	5•00
26•53 $ Monthly Payment	25•39	24•36
• • • • • • • • •	• • • • • • • • •	• • • • • • • • •
12•00 Mo.	12•00	12•00
168•55 Σ $ Int.	169•18	149•85
870•55 Bal. Due	864•50	857•53
• • • • • • • • •	• • • • • • • • •	• • • • • • • • •
24•00	24•00	24•00
348•72	311•85	273•16
712•72	702•49	690•52
• • • • • • • • •	• • • • • • • • •	• • • • • • • • •
36•00	36•00	36•00
474•25	422•84	371•69
520•25	508•80	494•73
• • • • • • • • •	• • • • • • • • •	• • • • • • • • •
48•00	48•00	48•00
557•56	495•93	434•50
285•56	277•21	265•22
• • • • • • • • •	• • • • • • • • •	• • • • • • • • •
60•00	60•00	60•00
589•37	523•72	457•78
−0•62	0•32	−3•81
• • • • • • • • •	• • • • • • • • •	• • • • • • • • •

the engineer attempts to obtain a business loan, he must go to the bank and fill out a series of nearly endless forms listing the assets and the liabilities of the business. The banker views this type of loan much differently, since he has a great deal more difficulty collecting it in the case where the business is unable to repay. As a sole proprietor or in most cases even as a partner, you remain responsible for the debts of the business; however, collecting them might require putting the business through bankruptcy proceedings and an endless amount of difficulty. In general, the private practice does not have enough attachable physical assets to reassure the banker. After you go to a great deal of effort in filling out the forms, and so forth, this type of loan is usually denied.

On the other hand, the private practitioner usually is a reasonably successful person and will have a fully developed line of credit. It is usually very simple to have one's *personal* credit extended to levels sufficient for the limited needs of the practice. This personal credit should be used carefully to assure that it is fully developed before starting the practice, since it will be required at one time or another.

One approach that can be of considerable help in record keeping is to use one particular card or line of credit *exclusively* for business purposes and use another with an entirely different firm for private and family charges. Then when the bills come in, there is no question about which are the responsibility of the business and which are personal and not to be entered into the business records. If this policy is rigorously adhered to, you will be thankful for the advice each time the income tax returns must be prepared.

Probably the worst aspect of the use of credit when engaged in private practice is the fact that bills come in every month with great regularity, but income is pulse-modulated in leaps and spurts. If one has two or three "dry" months, the regular, fixed charges for loans can be quite a burden. It is not unusual for the consultant to have invoices out totaling a quarter of his year's income and find himself looking at an empty bank account when the bills arrive. For this reason, I would summarize my advice on credit as follows:

USE CREDIT SPARINGLY.

USE CREDIT ONLY FOR BUSINESS PURPOSES, NEVER FOR PERSONAL PURPOSES.

KEEP THE PAYMENTS DOWN TO A LEVEL WHERE A THREE- OR FOUR-MONTH PERIOD OF SLOW COLLECTIONS WILL NOT BANKRUPT YOU.

10

Report Writing

The single task that the consulting engineer is called upon to perform most often is the preparation of an engineering report. Frequently this report is called out as a specific contract item. It is not unusual to find that a lengthy project will call for monthly letter reports and a final engineering report at the end of the project. Quite often the report is the only deliverable item required of the consulting engineer.

The report is an important part of your work in a great many ways, but it is particularly important in the propagation of your practice. There are several reasons for this:

1. After a consulting task is finished, the only real traces that remain behind are the receipt for your invoice and the engineering report(s). The people with whom you worked may leave the company or be reassigned to other duties. The personal contact and the mutual decisions were great, but the thing that the successors deal with is the report. It will provide their view of you and of what was accomplished.

2. The report very seldom stops in one location. In a large corporation, it will be copied extensively and sent to people who did not work with you and to other divisions. In a multinational corporation copies of the report will turn up in other nations. I have seen copies of a report written in Africa show up in places as diverse as Fort Wayne, Indiana and Stockholm, Sweden.

For good or ill, the report is a marketing tool that can either bring you repeat business or effectively shut it off. For this reason alone it is necessary that your reports serve their purpose well.

There are a number of textbooks on the market today dealing with the topic of effective engineering report writing. There are also a great many more books available on the effective use of the English language. It is not my purpose to expand upon these efforts. Instead, I would like to concentrate upon the things that seem to have made certain reports successful and other fail. For discussions of grammar, sentence structure and literary style, I would direct you to search elsewhere. This effort is particularly directed at the contents of the report and the effective organization thereof.

It is often said that a good newspaper article always contains the answers to the following questions:

> Who?
> What?
> When?
> Where?
> Why?

Quite often the answers are found in the first sentences of the article; for example:

MOTORIST INJURED IN SLEET STORM

Mrs. Marjory Wilson of Skaneateles was injured in a one-car collision at about 9:30 last night on Rte 20 at Pompey Rd. Her car skidded into a telegraph pole. Icy pavement was given as the cause of the mishap. . . .

Thereafter, the article will give details of the accident and note that a sleet storm was in progress, that Mrs. Wilson was proceeding West, etc. The article might continue with comment about the adequacy of county road maintenance, and so on.

The point is that the format is designed to give the reader some of the basic facts of the situation to attract his interest. He might thereafter read the remainder of the article. The exact handling of the opening is quite important to the interpretation the reader places on the remainder of the text. For example, suppose the editor has a real ax to grind about road maintenance. The headline could be written:

SLUGGISH ROAD MAINTENANCE
INJURES MOTHER OF THREE!

or somewhat more calmly:

ICY ROADS CAUSE CRASH

The headline sets a completely different tone for the article in each case.

Naturally, not all newspaper articles are written in this precise format, nor are all engineering reports written in a single format. Furthermore, the emphasis placed upon the various points in an engineering report is somewhat different. Engineering reports generally do not emphasize the "who" as much as the "what." Another consideration in most engineering reports is that a recommendation for future action is generally presented. When people ask an engineer to look at a problem, they usually want to know how he would go about solving it. For this reason, certain formats will contain a section entitled CONCLUSIONS AND RECOMMENDATIONS.

There are three principal formats in common use in engineering reports. These formats are called by a variety of names, but for our purposes we shall refer to them as the narrative, the formal and the Borden formula. We shall discuss the narrative first because it is the shortest.

THE NARRATIVE FORMAT

The narrative format is just what the name implies—a story-type recounting of facts and recommendations. The narrative format is usable chiefly in places where a very short report is required. It is particularly advantageous when a report must be supplied to a nontechnical manager describing what you have done. If some technical matter is required with the report, it is often appended so that it need not intrude upon the nontechnical reader. An example of a brief narrative report follows:

LETTERHEAD

July 12, 19XX

Mr. Fred Boomer, General Manager
XYZ Industries
. . . .
. . . .

Subject: Impedance measurements on the ASR-54 Antenna

Dear Mr. Boomer:

Following your telephone contact of July 1, an appointment was set up for July 5 to have me examine the measurement procedure used by your test department in measurement of the ASR-54 antenna and to determine, if possible, the source of the discrepancy between your measurements and the published data for this device.

On the morning of July 5, I was greeted by Mr. J. Jones of your test department and given a tour of the antenna range. Mr. Jones also demonstrated that the Smith Chart Plots of the ASR-54 antenna were at variance with the published specification data.

Upon an examination of the GP-6900 Impedance Measurement Set specified for these measurements I discovered that the procedure for compensating for feed cable length had not been followed in the initial calibration of the instrument.

After lunch I instructed Mr. Jones and his assistant Mr. Able in the technique for preparation of a suitable compensating cable. This cable was prepared and the ASR-54 antenna retested. In this case the antenna was shown to be within the published antenna specifications.

Since the preparation of a suitable compensating cable is unfamiliar to people not used to making radio frequency measruements, I have prepared a separate procedure instruction which is appended to this letter for the use of your test department. There should be no further recurrence of this problem.

Thank you for the opportunity to serve the XYZ Corporation.

Sincerely:

Thomas E. Mitchel

Note some of the features of the report. First of all, it is in chronological order. The first paragraph describes the initial contact and the problem as stated to the consultant. The second paragraph verifies the fact that the problem did appear to the consultant as stated. The third paragraph describes the relatively simple diagnosis of the problem, and the fourth presents the solution with a factual verification. The fifth paragraph contains the recommendation, namely: use the procedure prepared for calibrating the instrument and the problem will not recur.

The relationship between the second and the first paragraphs is important. If the initial contact with the consultant is made by a nontechnical person, the problem may not appear to the consultant as it was described to him. Often it is necessary to spend a considerable portion of the effort on a short job such as this in getting the client to realize what the problem is. Without this understanding, he may feel that you have solved the wrong problem.

A third item is a bit more subtle. You will note that there is no mention of the charges for this effort. Those will go on the invoice, also appended. The invoice will probably be for two days' work—the day chronologically accounted for in the report, and one day for preparation of the report and procedure instruction for use by the test department. The time will be spelled out in the invoice. On short tasks like this one it is often difficult for the manager to realize that you actually spent a full day at the plant. He saw you in the morning, and the next morning the problem was gone. The report notes both the chronological elapsed time and the fact that a separate effort in preparation of the procedure was required.

The narrative report is actually suited only to such short forms as the one presented. If it were to run longer than two pages, it would become clumsy and would not get the correct points across.

THE FORMAL REPORT FORMAT

When a task of any significant proportions is undertaken, it is usually advisable to follow a formal report format. This format is of the form:

TITLE PAGE
SUMMARY
INTRODUCTION
DETAILED FACTUAL DATA
CONCLUSIONS AND RECOMMENDATIONS

DESIGN ANALYSIS

of the

ASR-54 ANTENNA

prepared for the

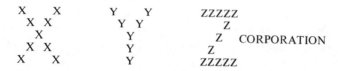

X   X   Y   Y   ZZZZZ
 X X   Y Y   Z
   X   Y   Z   CORPORATION
 X X   Y   Z
X   X   Y   ZZZZZ

12 Main Street

Auburn, N.Y.

14325

Note: Their name and logo
         larger than yours.

by

THOMAS E. MITCHEL

RF development

128 Packard Dr.
Syracuse, N.Y. 14397

Thomas E. Mitchel

7/12/77

Fig. 10.1. Sample title page.

The Title Page

The title page is a most significant part of the report, since it is the first piece to be seen. The place of honor on the title page should be occupied by the title, not by the name and logo of the client and certainly not by *your* name and logo. It should carry those items, along with the date and a signature, but they should be beneath the title. A sample title page is shown in Figure 10-1.

If possible, by the use of large type or some other device it is desirable to have the title itself dominate the top of the cover. The name of the client should be well above the name of the consultant, and DO NOT MAKE YOUR LOGO LARGER THAN HIS! That may sound silly, but I have heard clients complain about this particular thing.

As noted earlier, it is not a bad idea to have your printer prepare some sheets with the logo in the miniature size used on the envelopes, for use as report cover sheets. (See Chapter 8.)

The Summary

The summary is also an important part of the report. It should be placed immediately inside the title sheet. In a technical paper, the summary should tell a casual reader enough about what is in the paper to give him an opinion about whether it is worth his while to read it. In an engineering report, the summary may be the *only* part read by the General Manager, the Manager of Marketing, and so on. These people will probably have to approve any further work from you, and this may be your only shot at them. Make it count!

Suppose that the problem with the ASR-54 antenna had been a little more pernicious than in the previous example. A sample summary might read:

SUMMARY

Owing to difficulties in obtaining satisfactory impedance performance on the ASR-54 antenna, a design analysis program was conducted. Three errors in the drawings were found:

1. Incorrect feedhorn dimension.
2. Incorrect feedhorn placement.

3. Incorrect horn cover material used.

With these corrections a sample run of six antennas have been qualified within the impedance specification. The corrections have been added to the drawings.

Note that the sample summary tells *why* the program was instituted, *what* the program found and *what was done* to correct the problem. It is short and "punchy" to the point of being nearly telegraphic in style. It is also relatively nontechnical for the benefit of those within the organization who are not concerned with technical details but are concerned with the outcome of the program as a whole. The least-technical general manager should be able to read this summary and glean the fact that the problem has been corrected and the antennas are now salable.

In passing you will note also that the summary is placed upon a separate sheet with the margins much wider than the rest of the text. This is done deliberately to attract the attention of page flippers.

The Introduction

The introduction is that portion of the report in which you have an opportunity to give the background of the case. Once a problem is solved, people may have a tendency to forget that it ever existed, and they may wonder why they ever bothered to call in that dude from outside and pay him all that money. After all, they have managed to ship 465 of the darn antennas without a single hitch! This is your opportunity to document the situation as it was when you found it. The introduction should *not* be a horn-blowing exercise, but it *should* record the facts that led up to the work described in the report.

The use of a partially narrative style is common in the introduction, and portions of it will be roughly chronological. To continue with our example:

DESIGN ANALYSIS
of the
ASR-54 ANTENNA

INTRODUCTION

The ASR-54 antenna is employed with the airport surveillance radar being manufactured by the XYZ Corporation at their Houghton, Michigan plant. The antenna itself is being manufactured at the Auburn, New York facility. This radar was initially developed by the ABC corporation of Anaheim, California under contract to the FAA.

The initial items of this antenna, serial numbers 100, 101 and 102 were completed about February 15, 1977. In the course of testing in accordance with FAA-5438-76 para. 3.2.1.1 it was found that the antennas failed the impedance specification. (It should be noted that the ASR-54 is a "chirped" radar, and the antenna impedance must be controlled within tight tolerances. This is in distinction to the more usual antenna specification where VSWR is controlled and impedance allowed to vary.) After a mechanical inspection it was found that the antennas were built within drawing tolerances, and no mechanical departure from drawings could be detected.

In an attempt to resolve the source of this difficulty, purchase order 815-22-3-77 dated March 4, 1977 was issued for a brief appraisal of the measurement procedure. The result of this investigation was submitted in an engineering report dated April 1, 1977. The summary of the report indicated that the measurement procedure and the instrumentation were in order. It was suggested that the drawings might be in error.

In response to this assessment, purchase order 815-54-5-77 dated June 3, 1977 was issued for the present study. A detailed discussion of the effort follows.

Notice that the telegraphic style has vanished. The account is condensed but chronological. It attempts to point out gently that the first three antennas were unsalable (even though they would have passed the normal VSWR test used by XYZ to qualify antennas) because of the special nature of the radar that uses this antenna. In addition, it points out that the antennas were actually built to the drawings furnished by the customer, but this is of no help because the performance requirements always override. Anyone of normal perception should be able to read this introduction and realize that the Auburn division was in deep trouble by the third of June, and there should be no question of *why* the study was funded!

You will note that both the purchase order number and the date are given, despite the fact that a rudimentary date is included in the purchase order number. This is advisable because there is usually a chronological record of purchase orders as well as the serial file. The search can sometimes be quicker if both are listed, since a full month's worth of records need not be searched. I do not know why Purchasing loses these things, but they often do.

Another thing to note is that the introduction is somewhat more technical than the summary was; however, it is still couched in layman's language to the greatest extent possible. There is a good reason for this. Firms that physically build antennas like the ASR-54 in production are mostly hardware builders populated with mechanical engineers. They are building with a bunch of metal and plastic pieces that fit together to make an antenna. The skills are primarily mechanical. It is important that the introduction and the recommendations and conclusions be understandable to them.

Detailed Factual Data

This section should contain a fairly detailed account of the actual measurements, as well as analysis, corrective actions and test results of the program. While it is a good idea *always* to keep the technical level within reach, this part of the report should be technical enough to tell what was found and what was done to find it. In this section it is important that another engineer skilled in antennas (in this case) be able to follow your reasoning and check your work at some later date. If the marketing manager cannot follow this section, that is too bad. It would be nice if he could, but your goal here should be rigor rather than popular readership. Because of its technical nature no example will be offered. BE AS CLEAR AS POSSIBLE BUT DON'T OVER SIMPLIFY.

Conclusions and Recommendations

This portion should again be brief and forceful, but it must be precise as well. The various readers should know what you have concluded and what you have recommended even if the topic of the study is outside their field of expertise. Continuing with our example:

CONCLUSIONS AND RECOMMENDATIONS

The dimensional study of the reflector revealed that the feedhorn was 1.25 inches inside (toward the reflector) of the geometric focus of the reflector. This had apparently been brought about by the addition of the secondary gas barrier, which is the same thickness.

The error could have been corrected by reducing the length of the throat of the feedhorn by 1.25 inches. This would have been the preferable correction; however, it apparently was not done on the prototype built by ABC Corp. The correction employed apparently consisted of a 1.25-inch extension of the "J"-shaped feed waveguide, since this yielded the closest approximation of the original impedance data.

The horn cover material specified in drawing B-111-3467 is G7-A3 grade. This material is unsuitable for outside exposure and should be replaced with G10 A-3 material.

It is recommended that an engineering change proposal be instituted covering the 1.25-inch extension of the "J" feed waveguide, drawing number C-111-3461, to eliminate the use of the shim section employed to correct the first six units.

A second ECP should be instituted to cover the horn cover material change.

With these changes, it has been shown that no difficulty should be experienced in passing the impedance specifications of FAA-5438-76 para. 3.2.1.1

The formal-format report is not an example of the world's greatest literary style; however, it solves several problems for the consultant:

1. It recognizes the fact that the majority of the people who will receive the report will not read the whole thing. It flags the important parts for those who will browse but have to know.
2. It provides an orderly way of finding all of the facts of the situation.

THE BORDEN FORMULA

The Borden Formula represents a technique or organization that was taught for many years in the General Electric Company course entitled "Effective Presentation." This course was designed to cover both written and oral presentation, and ranged from formal reporting to letter writing and extemporaneous speaking. The technique is simple and widely used. It is one of those inventions which are so ubiquitous that few people realize that they were ever invented.

The basic parts of a presentation are broken up into the following sections:

HO-HUM
WHY BRING THAT UP
FOR INSTANCE
SO WHAT

Let us take a brief look at what these headings mean.

The "Ho-Hum" crasher is the opening few lines or paragraph. It is intended to attract the attention of the reader or listener, to wake him up. The "Why Bring That Up" is the section giving the reason for the communication in the first place. The "For Instance" is the main body of the argument. The "So What" conveys what the speaker or writer would like to have done. This technique has been used in literary endeavors for quite some time. We shall use as an example a short speech, which might strike a familiar chord. The editing and comments are mine—with apologies!

(HO-HUM)

Friends, Romans, countrymen, lend me your ears!
I come to bury Caesar not to praise him.
The evil that men do lives after them,
The good is oft interred with their bones;
So let it be with Caesar.

(WHY BRING THAT UP)

The noble Brutus
Hath told you Caesar was ambitious;
If it were so, it was a grievous fault,
And grievously hath Caesar answer'd it.

(FOR INSTANCE)

Here, under leave of Brutus and the rest—
For Brutus is an honourable man;
So are they all, all honourable men—
. . . .

. . . .
He hath brought many captives home to Rome,
Whose ransoms did the general coffers fill;
Did this in Caesar seem ambitious?
When that the poor have cried, Caesar hath wept:
Ambition should be made of sterner stuff:
Yet Brutus says he was ambitious;
And Brutus is an honourable man.
. . . .
. . . .

(SO WHAT)

I speak not to disprove what Brutus spoke,
But here I am to speak what I do know.
You all did love him once, not without cause;
What cause witholds you then to mourn for him?
O judgement! thou art fled to brutish beasts,
And men have lost their reason. Bear with me;
My heart is in the coffin there with Caesar,
And I must pause till it come back to me.

—Julius Caesar, Act III, Scene II

   Antony's funeral oration differs from engineering letters and reports in that it is steeped in irony and ends by saying one thing and implying something else. This is seldom required in an engineering presentation, but then engineers are rarely called to speak in front of a group of assassins who have just murdered the most powerful man in the world. Under certain circumstances, the prudent man doth his backside guard.
   The Borden Formula organization is useful in matters more mundane than earth-shaking speeches. It can enliven your letters, and make interesting reading of your chapters and subsections. Although it may seem a bit cheeky to provide a second example after having used Shakespeare for the first, the following letter is humbly submitted as an example of the use of the Borden Formula in a simple business communication. This letter example is written in a more familiar vein than the example of the narrative re-

port. This is not a feature of the Borden Formula but was done to illustrate that a business document need not be stiff and formal if the occasion warrants familiarity.

LETTERHEAD

May 10, 1977

Mr. Fred Boomer, General Manager
XYZ Industries
. . . .
. . . .

Dear Fred:

I want to reaffirm the conclusion of my April 1 report that your measurement procedures are in order and that the fault lies in the ASR-54 antenna itself. I further feel that the errors are not major and can be corrected fairly easily.

In response to your telephone call yesterday I would like to submit the following quotation for my consulting services in the investigation of this problem.

I feel that the investigation should concentrate on two principal areas:

1. The geometry of the reflector.
2. The location and performance of the feedhorn.

I would estimate that this effort would entail about four man-weeks of my effort spread over a period of five weeks. During this period the services of Jones and Able will be required full-time along with perhaps two man-weeks of engineering and drafting. I would estimate that three man-weeks of model shop labor will also be required.

The cost for my services for this effort will be 15 days $\times$ \$. . . plus vouchered travel expenses. This effort includes preparation of a final engineering report of approximately 25 pages. Three of the four weeks would probably be spent on your premises.

Terms and conditions are . . . .

I am relatively confident that the source of this discrepancy can be located within this period and corrective measures applied so that production can start rolling again. It will be a pleasure for me to serve XYZ in this matter.

Sincerely:

Thomas E. Mitchel

It might be interesting to go back to the beginning of this chapter and see whether you can identify the elements of the Borden formula in the structure. The writer uses it regularly!

11

Public Presentations

11
PUBLIC PRESENTATIONS

The consulting engineer is frequently called upon to make some form of public presentation. While this might not be quite as frequent a duty as the report, it is very nearly so. Despite the advice of Aesop, people *do* judge birds by their feathers and books by their covers. Your appearance and deportment at public presentations *will* have an effect upon your professional success. If you intend to succeed, it will be to your advantage to consider the impression that you make on clients.

When I refer to public presentation, I am not just referring to speeches delivered before learned societies or the school board. You may have to make a few of those; however, the number of those appearances will be relatively small compared to the business conferences and meetings which you will have to attend. These are the heart of the matter; your success at these business meetings and conferences will significantly influence your overall success.

The Business meetings come in three basic types. First is the pre-contract meeting. Before a project of any significant size is undertaken, you will generally be called in to discuss the matter with some of the client's people. During the course of a sizable project, there will be in-progress meetings to discuss the project and assess what is being accomplished. At the end of nearly any large project, there will be a conference to discuss the results of the program and the recommendations for continued effort, if such is required.   Since there are certain differences in these meetings, we shall discuss them separately and in order.

THE PRE-CONTRACT MEETING

The Development Program

Let us assume that the client company is about to embark upon some development program in which its in-house expertise is limited.   The company's management is considering the possibility of retaining a consultant to fill in the gaps in their skills.   In such cases they are likely to interview several consultants in an effort to select the best one for the project.

In these cases one probably will meet all of the major management personnel involved with the project.   They are all likely to show up for the interview conference in order to get a chance to size up the prospective consultant.   The gap in their skills is probably significant to the outcome of the development, or else they would not seriously consider the expense of retaining the consultant.   You will probably not be told whether or not they are considering others for the assignment, but it is usually safe to assume that they will if they can find someone who is qualified.

In this situation, the actual rate structure is of relatively little importance.   The client will be looking for experience, competence and a proven track record.   Many of the people at the conference will have little contact with the economics of the retention of a consultant and will care less.   They will, on the other hand, be quite concerned with obtaining the best possible outcome in the shortest feasible time.   Most will be well aware that a really competent person often can out-perform a less competent one on a development program.   The program that bogs down will invariably cost more money than the one that flies straight to the target, regardless of the rates that people charge.

The development pre-contract meeting usually commences with a detailed exposition of the position of the client, his goals and the requirement that brings you into the picture.   Often, there will be an old friend or two at the meeting; that is the way you were brought in.   After the exposition and some general comments from other attendees, it is your turn.

Now the ability to think on your feet is an invaluable asset. Before the meeting, your invitation probably contained some hint as to the nature of the requirement, and it is well to spend a bit of time beforehand reviewing what you know that is pertinent to it. But while doing your homework for background is valuable, the detailed exposition of the problem may contain elements that completely negate the value of your preconceived approach. Thus it may be necessary to operate entirely extemporaneously.

A gambit frequently helpful in such situations is to stand up at the board and make a *brief* list of the known ways in which their desired goal has been accomplished previously. You will also want to comment (again briefly) about the reasons, culled from their exposition, why each technique described is desirable or undesirable.

This approach has three advantages:

1. They may not be familiar with the various techniques and may see certain advantages to some.
2. It provides the attendees with an opportunity to interact with you. It may be that one of the techniques presented will spark a broader consideration of their requirements. This in turn could result in a better apportionment of the constraints placed upon the solution.
3. It both forces you to think about the situation in an orderly manner and provides you with some time in which to do it.

By the time this interchange is finished, you should have one or two preliminary ideas to toss into the pot that have some promise of satisfying most or all of the constraints. I usually feel it is important, at this stage, not to narrow the consideration too far. It is rare that all of the people present at the meeting will like the same approach to the problem. There are things to like and to dislike about any approach, and some of the preliminary schemes may turn out to have serious flaws when examined in detail. ABOVE ALL, DO NOT GET EMBROILED IN DEFENDING ANY PARTICULAR APPROACH AT THIS POINT. If someone does not like a particular approach, consider some others; they just might prove to be better in the long run! At the conclusion of such a meeting, if you leave the group with the feeling that you have a broad and solid grasp of the basic problem and have several reasonable approaches to the solution, that is about as well as you can expect to do.

Notice that I have suggested that you leave a meeting of this sort with several rather concrete suggestions about solutions. You will rarely be paid for these pre-contract meetings; therefore, the effort and your contribution constitute a "freebie." This is sometimes a matter of concern. Aren't they likely to take your good idea and use it to solve the problem themselves? Might they not suggest your good idea to another inverviewee? Well, I suppose that they might; however, in my experience, most of these people will deal with you in good faith. If they do not, those are the breaks of the game; and anyway good ideas are a dime a dozen. The pre-contract meeting is no place to get coy about ideas and proprietary rights. The client is placing his cards on the table too!

Problem Solving

The pre-contract meeting in the problem-solving situation is usually quite different from one related to development. In general, there will be far fewer people present. This may be simply a reflection of the old adage: "Success has a thousand fathers but failure is an orphan." You will usually be faced with a single manager or a manager and an engineer. The person whose recommendation brought you the invitation will rarely be present for the meeting, although you may meet him at lunch along with the General Manager. There is also likely to be a difference in attitude, which will prompt a difference in your behavior.

The chief difference between the problem-solving situation and the development situation stems from the fact that the former represents a problem to the person who is interviewing you. That person might be responsible for manufacturing the antennas which do not meet specifications and cannot be sold. The division is in trouble, the department is in trouble, and he himself is in trouble. He may have been directed to "get in an expert and solve this situation P.D.Q." and he may be concerned for his own job or at least his record. This is in contrast to the development situation, where the group was embarking on an adventure with high hearts. IN

THE PROBLEM-SOLVING SITUATION IT IS NECESSARY TO RECOGNIZE THE GRAVITY OF THE CASE AND TO RE-ASSURE THE PERSON WHO IS INTERVIEWING YOU.

Another difference exists in the urgency of the matter. Because of this urgency, the people involved are far less likely to shop around. Something or someone recommended you, and if you can show them some reasonable hope of success, they are anxious to get on with it. It is important that you share this sense of urgency with the people involved.

In the problem-solving situation there will always be something physical, whether hardware or software, involved. Because the group is smaller, you will usually be shown this hardware with a demonstration of the problem. You will usually be given the opportunity for a little hands-on manipulation of the measurement. With a single exception, cited below, IF YOU CAN SOLVE THE PROBLEM AT THIS POINT, DO SO! Even though you have no contract for the effort, a purchase order can usually be written to cover your day's effort and you will have left a group of happy, relieved and impressed clients in your wake. Few consultants experience any difficulty having their invoice honored in such a case.

The exception is perhaps best explained in terms of an example out of personal experience: During my late teens I was a RADAR repairman attached to one of the Army laboratories. An urgent request for assistance came in from an anti-aircraft battery assigned to protect one of our most vital military installations. It seemed that their semiautomatic-tracking gunlaying radar would pick up targets very easily but could not be made to track them. This left the guns without radar direction. World War II had ended, but the set was conspicuous to the brass, and heads were rolling.

A sergeant/pilot and I were ordered to see what could be done to put these folks on the road to success and happiness. We were met on the flight line by a captain who was the Battery Commander. The sergeant was ordered into the waiting staff car and I was ordered to attend to the airplane. With this the sergeant said, "Sir, the private is the RADAR man, I'm the pilot," whereupon the captain responded, "That —— kid!" (You can see we got off on the right foot.)

When we arrived at the RADAR site, the repairman-in-charge turned out to be a 28-year veteran sergeant who had been working in radio and radar throughout his career. He managed to donate a few more slurs about me personally and about the ancestry and IQ of the people at the lab. He further indicated that the failure of the set represented a clear case of demonic possession, and that if he couldn't fix it no —— kid would do it in a million years. He then graciously demonstrated the flaw in its performance.

I had slept little the previous night, going over the manuals for the set in an attempt to divine what might be wrong. It was obvious that nothing major was out of kilter, since the set easily picked up airplane targets. It then occurred to me that the problem could exist if the various indicators were being synced to start at entirely different times. A careful check showed that two co-axial cables entered the main indicator side by side. If they were interchanged, that would produce the same symptoms. I knew what I was looking for, and the first demonstration of the malfunction convinced me that I had made a lucky guess.

By that time, those jaspers had my dander up, and I had decided to show up the loud-mouthed sergeant or die trying. Before the sergeant had finished his demonstration, I stalked out of the trailer that housed the set, grabbed the ladder stowed beneath, opened the rear doors, switched the cables, closed the doors and replaced the ladder—all to the accompaniment of a steady stream of profanity from within. I then entered the set, found a new target and threw the set onto the semiauto track—whereupon it locked on and hung on! I had guessed right, and it was fixed!

At this point the sergeant became livid, shrieking artful embellishments upon his original description of my ancestry and character and stoutly maintaining that the whole thing was a trick. Nobody could fix the radar that fast, least of all your humble servant. At this point I blew my cool, retaliated in kind and then proceeded to repeat the performance in reverse order, leaving the set again malfunctioning. I then sat down, lit up my corncob and observed the tumult about me while trying to regain my composure.

The captain, who had been a spectator throughout this charade, joined the action with the comment: "Private, you get your —— back in there and fix that set and then you get your —— back to the lab. You are on Report!" And so I was! A very lenient Company Commander put me on company punishment for a week for discourtesy to superiors. It was my good fortune that the punishment was served in the relaxed atmosphere of the casual company housing the lab people rather than that AA battery!

Hence the exception that was the original stimulus for the narrative: NEVER, EVER, UNDER ANY CIRCUMSTANCES DO SOMETHING TO DELIBERATELY SHOW UP ANYONE!

People involved in problem solving are bound to be concerned and sometimes fearful. They may even be a bit less than courteous. They may be entitled to be. If you can stay calm and controlled, you can usually bring them around. ALWAYS ALLOW THEM TO PARTICIPATE IN THE PROBLEM-SOLVING EFFORT. This saves their "face" and costs you little. It may win you lifelong support.

The cases where one encounters outright hostility are not common, and they seem to occur in inverse proportion to the number of gray hairs at your temples. Whether this phenomenon is due to superior maturity on the part of the consultant or is in deference to age is the topic of another study.

THE IN-PROGRESS PRESENTATION

An in-progress meeting occurs when various people within the client organization would like to find out about what has been accomplished so far. These meetings generally are called only for projects of substantial length. If there is an in-progress meeting for a problem-solving type of project it is usually because the problem has not been solved; and the consultant is now in trouble along with management. In the development-program situation, an in-progress meeting is probably a prescheduled affair, due simply to the length of the program. Some of these meetings will consist of a simple ticking-off of the various milestones which have been passed. Unfortunately, others will have to deal with things that have been tried and proved not to work.

The most pressing problem to confront the consultant at the in-progress meeting is that of what to say and how to comport himself at the meeting when he is forced to tell the client that something did not work. The brightest people will try things that do not work, and at times you may see a program fail in which you had what seemed to be the clearest of insights. Edison observed that invention was 1% inspiration and 99% perspiration. He freely admitted that he had failed far more often than he had succeeded, that success was far more often a matter of persistence than of sheer genius.

This problem is compounded by the fact that a long program tends to wear down the enthusiasm of all but the heartiest proponents. It is not unusual to find that a program originally forecast as an 18-month effort is subjected to heavy management fire at the 6-month mark if early results are negative. This is like the anecdote about the grizzled, battle-scarred veteran of nine years' service shambling into a recruiting office and saying: "Hey Sarge, could you give me that talk again? I'm losing my enthusiasm!"

From the viewpoint of management, a 6-month record of inadequate success is food for thought. So the program is supposed to run 18 months. SO WHAT? This dude has tried his best idea, and it didn't work or at least it didn't work well enough. Wouldn't we be better off to drop the whole idea and cut our losses?

There is actually a reasonable amount of sense in such a reaction. After all, not every development program succeeds. The business history of the United States is littered with Edsels and F-111's, which someone originally thought were great, but which did not work out. Why pour good money after bad?

The more sports-minded reader probably by now has a mental picture of Knute Rockne standing in the locker room at halftime. The team is down 21–0. He clears his throat and delivers an oration that sends the players out to victory. Stirring leadership and fighting hearts carry the day. That is great emotional material, but it isn't business.

When the early efforts in a development program fail, there comes a time for scrupulous honesty on the part of the consultant. By nature, and in order to survive, he must be possessed of a fighting spirit and not be prone to give up too easily. Tenacity has a great survival value in this business. If you are convinced that you can succeed within the constraints of time, money and materials, this is the time for a demonstration of confidence and leadership.

These qualities must be accompanied by some solid plan for attaining success.

Few of the major advances made in our society would have come to fruition if someone had not been willing to fight off midpoint failures or obstacles. People who have not participated in the development of something new rarely appreciate this fact. They may require your spirit and determination to lead them.

If, on the other hand, it seems that the program is aimed too high and that some of the original goals are not within reach, now is the time to say so. Do not waste their money and time unless you can see some way to succeed.

A development program that brings in a consultant can succeed only if the consultant can fight through early setbacks. HAVE A PLAN FOR SUCCESS AT ALL TIMES. KEEP YOUR SPIRITS AND CONFIDENCE UP.

THE FINAL REPORT MEETING

At the close of nearly any development that has met with any reasonable degree of success, there will usually be a final report meeting. The purpose of this meeting will be to review the final engineering report and evaluate what the program has accomplished. Such meetings are generally happy occasions, since some or most of the goals have been accomplished. If they had not been, the program simply would have been terminated, and no one would be interested in having a meeting. At such a meeting you should be prepared to go through the program systematically and compare what has been accomplished with the original goals. A format that is useful in many cases is to follow through the original statement of work, item by item, and summarize the outcome of each individual task. This procedure has the advantage of clearly spelling out to the client the fact that he has received what he has paid for.

After this point-by-point evaluation, it is often worthwhile to follow with conclusions and recommendations. Here you should summarize anything that is novel and potentially patentable or of proprietary advantage. It will not always be apparent to the client that certain features of the development are novel and possibly protectable; therefore, it is the duty of the consultant to point them out.

It is also worthwhile to point out any other uses of the information and techniques developed in the program. During the course of the program you will have seen a great deal of the client's organization, and it is the rare case in which a new device or technique does not have more than one useful application. It is the duty of the consultant to point these applications out, and it is good business besides!

There are nearly always following steps to be recommended. In order to make use of what was developed, one should next do A . . . , B . . . and C . . . . It is certainly valid to bring out these next required steps. However, DO NOT PERMIT A HARD-SELL PITCH FOR MORE CONSULTING SERVICES TO ENTER INTO THE FINAL MEETING.

The client has bought and paid for what he has, and he should be able to use, neglect or dispose of it as he sees fit. The purpose of the final report meeting is to tell him what he bought and how he can use it. SAVE THE SELLING FOR LATER!

ATTITUDE

The various attitudes that are beneficial to the consultant have been discussed at length, in connection with various detailed points of application; therefore, I shall not add much here. There is probably nothing required of the consultant other than good manners and an observance of the golden rule. A fair, even, honest and conscientious way of dealing with people will go far in this business.

One particular point should be emphasized. In order to succeed, the consultant will generally have to be a bit brighter than the majority of his fellow engineers. People have to respect him in order to spend the money and make the effort required to bring him in. When he is brought into a situation, it is usually in a position of honor. It is therefore important that he make every effort to avoid being a smart aleck and being obnoxious and objectionable to new associates. A military officer and a General Manager can get along on respect alone, but a consultant needs to be respected *and* liked in order to succeed.

DRESS

A recent best-selling book by John T. Molloy is entitled *Dress for Success.* The basic tenet of the book is that clothes *do* make the man and the woman. Over a period of 15 years, Molloy collected data from over 15,000 executives across the country and came to the conclusion that successful people look successful.

It is interesting to note that the book itself appeared in two different covers. One showed a man's dark blue pin-striped suit with a white shirt collar and a pale blue tie, with the title printed in white letters. The second cover showed a very loud tie with a suit, and the printing was in red. The first cover outsold the second at the retail level in Rochester by about 5 : 1. People in the retail book business will assure you that people *do* judge a book by the cover.

In a recent issue of the magazine *Dynamic Years*, Paddy Calistr published an article entitled "How to Dress for Business." The author checked Molloy's conclusions by contacting the personnel departments at General Motors, Hallmark Cards, Coca Cola and a variety of other businesses. It was found that, indeed, dress codes called for attractive and generally conservative attire at nearly every large corporation. Anyone dealing with the public would not wear jeans, T-shirts, love beads or excessively long hair.

For generations people have believed that one should dress his part in life. Soldiers should look like soldiers, and airline captains should look like airline captains. How secure would you feel on a flight if the captain showed up in a pair of bib overalls and sandals with love beads around his neck?

The main problem is that no one seems to know what a consulting engineer is supposed to look like. I imagine that a cartoonist who had to represent a consulting engineer might draw someone in academic robes wearing a mortarboard with a lightbulb on top. I once saw a consulting aerodynamicist show up at a meeting at Wright Field wearing a pair of pajama bottoms tied with a string plus an undershirt surmounted by a long scraggly beard. Somehow, neither of these getups seems appropriate to me.

When our esteemed treasurer discovered that I would write a few notes about style into this text, she shook with prolonged uncontrollable laughter. If I were called upon to classify my own style, it would come out something like "midwestern-stodgy." I am still wearing the crew-cut hair acquired in high school. I am given to bow ties, short-sleeved shirts and athletic socks. The crew cut does not blow in the wind and requires only the services of a towel when emerging from swimming; the bow tie is much safer when working around rotating machinery; and the athletic socks are a great help in the prevention of jungle rot and athlete's foot. I must admit that the short-sleeved shirts are a bow to comfort in favor of style. Molloy maintains that the would-be executive never, ever, wears a short-sleeved shirt, since it is an indelible mark of the lower middle class. However, it is a bit of a nuisance to be forevermore rolling up sleeves before you can do any work. They do, of course, offer the additional advantage of hiding tattoos on the forearms, should you have any.

Taken as a group, engineers are not at the forefront of style. Shirtsleeves are the rule rather than the exception in an engineering office. I used to have a boss who said that whenever he saw anyone in an engineering office with his suitcoat on, he knew that he was charging his time to overhead rather than a paying project.

Probably the best thing to remember is that a consulting engineer is forevermore selling. It is probably not a bad idea to get your clues on dress from the salesmen who come to call on you. They are not so fashionable as to be mistaken for a hairdresser who wandered in by mistake and not so poorly dressed as to be mistaken for the janitor. If you show up at that pre-contract meeting in patched jeans, sneakers and a Harvey Wallbanger T-shirt, don't be too surprised if they decide on some other consultant.

12

Do I Really Want To Do This?

When I am out on the road, after completing a long working day, one of the client's engineers and I frequently repair to some local restaurant for dinner. It is usually some rather nice place, selected as much for the atmosphere as for the cuisine. We are anxious to leave the thoughts of work behind and will intentionally turn the conversation to other topics. The process of struggling together on a difficult task very rapidly builds a bond of friendship between people, and the conversation often turns to families and personal hopes and aspirations.

Sometime after the first cocktail has been ordered, my companion is likely to disclose that he has some idea for an invention or service that he has been considering. The idea rarely turns out to be half-baked. It is often sound and well thought out, the reason for this being that a client firm usually has its best people work with a consultant. Clients naturally want to maximize the return from their investment.

Throughout dinner, this idea will be batted back and forth, interspersed with anecdotes and reminiscenses. By the time dinner is finished, the topic will have changed.

If we have worked together for some time and feel that we know one another reasonably well, I will sometimes hear a final question: "Jack, what do you think? Should I give it a try?" I have a pat answer: "If nothing I can say or threaten you with will talk you out of it, you will probably make a go of the deal."

The answer is supposed to be humerous of course; however, there is a certain ring of truth to it. A person may *have* to be very determined to make a go of a practice or a small technological business. If he is that determined, he has one of the basic qualifications for success. And, in fact, nothing I or anyone else could say *would* talk him out of it.

I am not about to suggest that everyone who goes into these things and tries to make it on his own will have the fierce drive of a Charles Lindbergh. However, among the people whom I have known to succeed, a very strong desire to be self-employed and self-determined is a common characteristic. Several other personality and intellectual traits are also common among those who make the grade.

THE SELF-STARTER

Probably the first and foremost of the characteristics required by the self-employed engineer or technologist is that he be equipped with a pretty good self-starting mechanism. This attribute is frequently evidenced in the way the person conducts himself while in the employ of others.

If you stop to consider the population of your office or your place of business, I am sure that you can think of people who seem always to be busy doing something to enhance the company, even when there is not a great deal to be done. I am not talking about the person who manages to "frantic about" without accomplishing anything so that the assigned work always manages to fill the available time. I am instead talking about the person who always seems to finish assignments ahead of schedule and then suggests to the boss that the time left over may be applied to doing something that needs doing.

Often such a person will not seem frantically busy to less observant companions. He probably takes coffee breaks and swaps stories along with the rest of the group, but his work always gets finished ahead of time. Interestingly also, he is very often not the person who is in the plant until all hours every evening. There is some voluntary overtime, when the job at hand calls for it, but it is not a continual practice.

No doubt there are certain jobs where the regular working day does not suffice to see the work accomplished. Certain people are continually called upon to do more than they can possibly do within office hours. Furthermore, some of these people are enormously productive. On the other hand, the habitual overtimer may just be a putterer who takes longer than most to get his work done.

The putterer is probably the worst possible risk for self-employment. The new on-your-owner will invariably be called upon to work harder than he has ever worked in his life and will not have the added discipline of a management schedule to meet. In general, the putterer will get worse rather than better.

The workaholic, on the other hand, may well find that the added demands of self-employment will still further commit his resources. He can easily progress to the breaking point, where his health and composure may suffer. More than one person has *physically* failed under the stress of self-employment or entrepreneurship. He is a far better risk than the putterer from a financial point of view just because he is used to working hard. However, he also has developed the habit of doing an inordinate amount more than his fellows, which could be his undoing as a consultant.

Far and away the best risk for self-employment is the person who has always managed to get his work done a bit ahead of time, but has not been a chronic overtimer.

There is another characteristic often associated with this last group of workers. Frequently those in this group have managed to be doing something outside their normal work in a fairly serious fashion, *without* having it reflect upon their employment. It may have been a hobby, or it may have been an entrepreneurial activity undertaken at least nominally for profit. Several of the entrepreneurs whom I have seen succeed are people who perpetually had something going on their own time while they were employed by others.

Of these two subgroups, the part-time entrepreneur is the better risk if he was able *consistently* to make some profit. The hobbyist is more likely to have a happy-go-lucky attitude, which could be carried over. It should be noted, however, that a great many ultimately successful garage businesses were started as hobbies.

Hewlett-Packard probably falls more into the category of part-time entrepreneurship. It was started as a part-time effort to capitalize upon skills not fully used. On the other hand, the Ford Motor Company was founded upon the hobby of an upper-middle-class chief engineer employed by the Detroit Edison Company. Both approaches can work.

A final word on the subject of self-starting has to do with the subject of discipline in rising and work habits. At the start of self-employment, business is often scarce. It sounds like a contradiction; but, in truth, this is an easy time to get into some bad habits. Without the discipline of a formal starting time it is easy to get into the habit of rising late and doing little. There are old sayings to the effect that the less there is to do, the less it is likely to get done. Conversely, it is said that if you want something to be done for sure, give the job to a busy person. It is at this early slack period that the newcomer should be making the greatest efforts to put the business in order and to make new contacts and establish new lines of communication. Because of the normal delays in getting anything new started, a lot of time will be spent waiting. It is important that the newcomer find something at least moderately productive to do so that he does not just *sit* and wait.

This free time can be employed in getting the facilities into better order or perhaps put toward some task or development that would be marginal if more paying work were present. In the latter case, there is something to show the potential customer or client when he comes to visit. WHATEVER YOU DO, DON'T LET THE GRASS GROW UNDER YOUR FEET!

RESOURCEFULNESS

A great deal of resourcefulness is usually required of the on-your-owner. In the beginning of the operation there will not be enough money for equipment, supplies, tools and a host of other things that will be needed. In order to succeed, it will be necessary that he be able to "make do" and to improvise in all manner of ways.

In visiting a newly started, small technological business, it is common to find that the instruments are partially old and surplus items that have been refurbished and partially homebrew items that have been designed and fabricated on the spot. When it is done carefully, items of test equipment and tools can be purchased for a fraction of the price of more modern equipment, particularly when they need some repair. Similarly, special-purpose equipment can be designed and fabricated rather inexpensively when labor is not included.

At the outset of the business, it is likely that the load of paying work will not be so great as to fill your time. During this period effort should be devoted to building up the facilities. Any time not spent in doing some paying work can be so used. It is a good idea to do a professional-looking job on the equipment, since you may wind up by having to show it to a client using it on his project. If the equipment is too haywire-looking, it will not enhance his confidence.

Naturally, this admonition varies with the type of business being started. A person intending to establish a consulting service to furnish computer software will not require anything like the inventory of a group intending to develop and market a new Ham transceiver, since the product of the software consultant will consist almost entirely of paper with things written on it. But there is still the matter of office and facilities, and anything that the business can do for itself will not have to be paid for.

The matter of resourcefulness also extends to the subject of obtaining paying work. A source of income that can often be tapped is the design, development and fabrication of special test equipment. One-shot test fixtures for checking out printed circuit boards or fixtures for testing or assembling hydraulic valves and similar items are frequently subcontracted by the equipment manufacturer. These items are not a part of his product proper; they are only used to test and check the product. For this reason, he will frequently be happy to pay for outside effort on these items so that his own engineering staff can concentrate on salable product design. This is the obverse side of the coin discussed in the previous paragraphs.

Subcontracting of test equipment, special-purpose interface equipment, jigs and fixtures, and so on, tends to be less profitable than actual consulting work or new product development. The reason for this is the attitude of the client or customer. Since these items are not salable, he will tend to view them as straight expenses, and the smaller the better. If you offer to build a black box to connect two pieces of test equipment, the price that can be obtained is probably controlled by the cost of the two boxes to be connected, almost without regard to the actual effort involved in producing the black box. Often, these tasks have to do with modernizing equipment.

For example, some time ago a client wanted to have a tensile test machine (which was a separate free-standing entity) connected to a small computer to automate the data-gathering process. The tensile test machine stretched specimens with a mechanical gear drive, which was geared also to the paper drive in a chart recorder. The stress or force from stretching the sample was measured with an electrical load cell, and the chart recorder pen was caused to track it. The result was a stress/strain curve for the material. The operator would then note the slope of the curve by counting on the graph the difference in stress and strain between two arbitrary points on the curve. The slope of the curve was obtained by dividing. A second point of interest was the point of inflection where the curve started to knee over. This was found by laying a ruler along the straight portion of the curve and noting the point of departure. The energy of rupture could be calculated by counting the number of squares under the curve. Obviously if very many specimens were to be tested in a day, this process would get to be pretty tedious for someone smart enough to be taught to do the arithmetic.

The stress or force measurement was already electrical. However, it was in the wrong format; that is, it was an analog voltage proportional to the force. This had to be converted to a digital format for the computer and required certain logic added so that a given force reading would be loaded only once.

In addition, there was the problem that the strain reading (the amount the sample had stretched) was taken by the machine in a purely mechanical fashion. No electrical signal existed anywhere that was proportional to the strain. A device would have to be built to pump out a digital signal proportional to the rotation of the lead screw on the machine. Since the screw ran through many turns and it was desirable to have a controllable spacing between readings, digital circuitry had to be provided to permit selection of the fineness of the grain of the readings and to keep track of the position of the screw.

The black box that did this contained only three printed circuit cards. Also there was the mechanical transducer which clamped on one of the shafts in the screw drive. Compared to the desk-top computer, which had cost $6,800, and the tensile test machine, which had cost about $8,000 quite a few years ago, the box was a pretty simple item. However, it did have to be designed, tested, guaranteed and serviced. On something produced in quantity, these costs can be defrayed over a large number of units; but when you are only going to build one, that one must bear all of the cost. In order to get the order approved, the cost of the unit would obviously have to be smaller than the cost of the computer, which was tremendously more complicated. Needless to say, this made the black box a relatively low-profit item for the amount of effort that went into it. Only rarely, when the customer needs the device very badly, can one obtain a substantial return for the design and development time involved in this sort of interface item.

It should be noted that had this been an item the client intended to *manufacture*, the story would have been entirely different. Here he would have been able to defray the cost over a large number of items, and a reasonable price for the development could have been charged.

Despite the drawbacks to this type of business, the field of custom engineering of jigs and fixtures is a worthwhile area to investigate, since it is relatively easy to obtain this type of one-shot work. When the business is new, this type of work can provide bread-and-butter income.

FRUGALITY

A simple review of the economic models of Chapter 6 will suffice to show why frugal people have a better chance of survival in this business. The point needs no further elaboration.

THE UNDERSTANDING SPOUSE

A point that requires careful consideration when the would-be on-your-owner is married is the question of the ability and strength of character of the spouse. If the demands placed upon the adventurer are large, the demands placed upon the mate are larger still. The demands are sufficiently great to place a substantial strain upon a marriage, and cases where a marriage has failed because of this strain are not unusual.

In the calculations of Chapter 6 we considered that Mr. High Average was married and had two children. There was no financial contribution made by the spouse in this example. It is obvious that if the spouse had had a good paying job, their life would have been considerably less trying during the third year. It was instead assumed that she (or he) maintained a traditional homemaker role. This does not under any circumstances mean that the load upon her was not considerable. Consider, for example, the fact that she was the one who had to make do with the very skimpy budget and offer support and comfort as the business grew only marginally fast enough to stave off bankruptcy, and the savings dwindled to near the vanishing point. In the real-life situation she had to make a great many sacrifices. There were times when a new piece of equipment costing hundreds or thousands of dollars had to take precedence over the replacement of a frayed overcoat or a worn rug or a hopelessly rusted automobile.

The case of the working spouse is also fraught with some peril. Suppose that she (or he) has some reasonably good paying position. During the pre–self-employment days she probably had some household help so that when she returned from work she was not faced with that further demand. At about the three-year point of your practice she could get pretty tired of doing without the help and having to return from work every evening only to be faced with housework.

Then there is the fact that her mate will be spending much more time working than he spent at any previous time in their marriage. Sixty to eighty hours per week is not an unusual amount of time at the three-year point—which usually boils down to the fact that he will not be able to spend nearly as much time with her or to offer even a fraction of the help he provided with the household in the pre-business days. Add to this the fact that he is liable to be grouchy from overwork and, to some extent, from self-doubt.

An additional factor that can enter into the equation is the increased travel that characterizes self-employment. When he travels, it will usually be at the expense of the client. He is more or less expected to stay at and eat in nice places. A return to a blizzard-bound home accompanied by a description of a week spent swimming in the sunshine and eating steak at the most wonderful restaurants is liable to produce a spirited response from a mate who has been snowbound, eating hot dogs, and has not had a vacation in three years!

These trials are by no means certain, but they seem to be fairly likely for most on-your-owners. Their possibility should at least be considered and discussed *before* you turn in your resignation.

SENSE OF HUMOR

The final but by no means least important property that seems to characterize those who succeed is a sense of humor and the ability to avoid taking oneself too seriously. As you have seen, self-employment will consume a pretty fair amount of the humor that you have—so you had better start out with a substantial supply or you might run out.

THE FUN OF IT

If the substance of this text has been discouraging, this stems from the desire to do what Howard Cosell describes as "telling it like it is." There is no question that starting and managing your own engineering practice is a difficult and trying proposition. However, if the problems are great, the rewards are great also. I am writing these finishing lines from the front porch of a summer camp overlooking one of the Finger Lakes. Shortly I shall be swimming. No one told me to start this book and no one gave me a deadline to finish it. Our esteemed treasurer points out that in seven years I have not once complained about management.

At a point in life where some of my friends are trying to hang onto a job for the next 10 to 15 years so that they can retire, I relish each new challenge for the excitement and fun and new friends it brings. In the last three years I have worked in Africa, Belgium,

Germany, Norway, Sweden and Denmark. I am currently hoping that a project in Venezuela will come my way. I have had the privilege of working on new, modern and interesting projects ranging from antenna couplers for supertankers to a control link for an automated magnetically/levitated railroad.

It has been worth every bit of the pain and strain.

GOOD LUCK!

Index